A NATURALIST'S GUIDE TO THE
MAMMALS OF AUSTRALIA

Peter Rowland and Chris Farrell

JOHN BEAUFOY PUBLISHING

Reprinted in 2025

This edition published in the United Kingdom and Australia in 2021 by John Beaufoy Publishing Ltd
11 Blenheim Court, 316 Woodstock Road, Oxford OX2 7NS, England
www.johnbeaufoy.com

Copyright © 2017, 2021 John Beaufoy Publishing Limited
Copyright in text © 2017, 2021 Peter Rowland and Chris Farrell
Copyright in line drawings © 2017, 2021 Thomas Rowland
Copyright in photographs © see below and as credited throughout
Copyright in maps © 2017, 2021 John Beaufoy Publishing Limited

Photo Credits
Front cover: *main image* Koala © Brett Jarrett; *bottom row, left to right* Short-beaked Echidna; Eastern Grey Kangaroo; Bare-nosed Wombat, all © Peter Rowland/Kape Images. **Back cover:** Eastern Quoll © Sharon Wormleaton. **Title page:** Common Brush-tailed Possum © Angus McNab. **Contents page:** Tasmanian Devil © Sharon Wormleaton.

All rights reserved. No part of this publication may be reproduced, stored in a retrieval system or transmitted in any form or by any means, electronic, mechanical, photocopying, recording or otherwise, without the prior written permission of the publishers.

Great care has been taken to maintain the accuracy of the information contained in this work. However, neither the publishers nor the author can be held responsible for any consequences arising from the use of the information contained therein.

ISBN 978-1-913679-07-1

Edited by Krystyna Mayer

Designed by Gulmohur Press, New Delhi

Printed and bound in Malaysia by Times Offset (M) Sdn. Bhd.

Contents

Introduction 4

Australian Mammals Past & Present 4

Australia's Vegetation 10

Using This Book 17

Glossary 18

Species Descriptions 20

Checklist of the Mammals of Australia 160

Further Information 171

Acknowledgements 172

Index 173

◼ Introduction ◼

INTRODUCTION

Australia is one of the most biologically diverse countries in the world, with more species of mammal than 93 per cent of other countries, and has more endemic land mammals than any other country. Many of these are internationally recognized, like the Koala and Red and Grey Kangaroos, while others, like the Platypus, mystified the scientific world when they were first sent to the museums of Britain. Yet the bulk of the mammals in Australia are seldom observed by the general public, scurrying through the undergrowth at night, flying through the skies once the sun goes down, or concealed for most of their day beneath the surface of the oceans – and some even dig holes in our precious lawns while we sleep. Some Australian mammals are tiny, weighing only a few grams and being about the same size as the teabag I am just putting in my cup, while others, such as the wider roaming ocean dwellers, include the largest animal ever to live, the Blue Whale, weighing up to 200 tonnes and measuring around 30m in length.

At the time of writing there are 351 species of native mammal recognized in Australia, and a further 30 classed as extinct. Many of Australia's remaining native mammal species are considered threatened, with a large number of those being Critically Endangered. Sadly, many of them are killed on roads, electrocuted by power lines, become trapped in barbed wire fences or get stranded on beaches, and it is only then that we tend to see them. Australia is also home to 33 introduced mammals, including ourselves, the humans.

Often the best way to see native mammals is to join a field research team that traps secretive animals which we would otherwise not get a chance to see, and it is researchers on such teams that provide us with the information, and many of the images, that make a book like this possible; a book that aims to showcase the wonderful mammals of Australia.

AUSTRALIAN MAMMALS, PAST AND PRESENT

Australia's mammal fauna has been shaped from around 40 million years of isolation from the other continents in the world. The monotremes and marsupials have a higher percentage of endemics than the bats and rodents, with the latter likely to be much more recent arrivals in Australia (perhaps as recent as 1 million years ago).

MEGAFAUNA

Well before the present day, as was the case on other continents, mammal megafauna roamed Australia during the Miocene, Pliocene and Pelistocene Epochs. Animals like *Procoptodon goliah* (a 200kg, 2m-tall kangaroo), *Zygomarturus trilobus* (a 500kg, 2.5m-long 'wombat-like' diprotodontid), *Thylacoleo carnifex* (a 160kg, 1.5m-long marsupial lion) and *Zaglossus hacketti* (a 30kg, 1m-tall echidna) would have roamed Australia along with the early Aboriginals. What caused their extinction is subject to much debate, with the latest evidence suggesting that many may have been hunted to extinction about 45,000 years ago, with the extinction of the remaining species linked to the last ice age, which occurred at the end of the Pleistocene, around 13,000 years ago.

Australian Mammals, Past & Present

MODERN-DAY MAMMALS

Following the most recent published taxonomy by Stephen Jackson and Colin Groves (see Further Information, p. 171), the 415 modern land and aquatic mammals recorded in Australia occupy 53 families.

Monotremes are peculiar among the mammals. They have all of the mammal characteristics. They are warm blooded, have hair and produce milk to suckle their young but, unlike any other mammal, they lay soft-shelled eggs from which their young hatch. Australia is home to the world's only platypus, and the only short-beaked species of echidna. The three species of long-beaked echidna are now considered to be restricted to New Guinea, since the Western Long-beaked Echidna *Zaglossus bruijnii* was declared extinct in Australia, with the last Australian specimen collected in 1901.

Dasyurids are insectivorous and carnivorous marsupials, which are characterized by teeth that are made for biting and cutting, with seven pairs of incisors (four pairs in the upper jaw and three pairs in the lower jaw), four pairs of well-developed upper and lower molars, and at least four non-fused toes on the hindfeet and five on the front feet. Australia is home to 60 species, which range from the world's smallest carnivorous marsupial, the tiny Long-tailed Planigale, measuring just 55mm in body length and 4g in weight, to the world-renowned and muscular Tasmanian Devil at up to 9kg and 650mm.

Bandicoots are timid, mainly nocturnal ground dwellers, and only nine species remain today. They are often confused with rodents, but are marsupials with backwards-facing pouches to prevent the developing young from getting a face-full of dirt. Some species of bandicoot have the shortest gestation period (pregnancy) of any marsupial, at just 11 days.

Wombats are the bulldozers of the Australian bush, equipped with a compact, stocky body, muscular limbs and strong claws for digging, a thickened bony plate in the rump for protection and a rear-facing pouch for protecting their young. Three species occur today, including the Northern and Southern Hairy-nosed Wombats, both listed as threatened (Northern is Critically Endangered and Southern is Near Threatened on the IUCN Red List), and the widely distributed and more common Bare-nosed Wombat.

Possums, gliders and **cuscuses** are a large assemblage of around 30 arboreal, mostly nocturnal marsupials, and include the pygmy-possums, striped possums, Leadbeater's Possum, lesser and greater gliders, ring-tailed possums, Honey Possum, feather-tailed gliders, cuscuses and brush-tailed possums. Some species readily enter houses and search through bins for food, with their diets mainly consisting of plant matter, including fruits, pollen, nectar and sap, but they also eat insects, birds' eggs and even nestling birds.

Bettongs and **potoroos** have been impacted greatly by feral animals, altered fire regimes and loss of suitable habitat, with only two-thirds of the group remaining today and four of

Australian Mammals, Past & Present

those eight species listed as threatened. They are ground-dwelling marsupials with large hindfeet, and can hop like kangaroos when evading danger, although they push off with their hindfeet and land on their forelegs when grazing.

Macropods are all relatively large marsupials, ranging from the Quokka (smallest) to the Red Kangaroo (largest), and 42 remain today (with a further four declared extinct). Macropods are a versatile and familiar group, and include the wallabies, the open plains-dwelling Red Kangaroo, the more scrub-dwelling grey kangaroos, the wallaroos and rock-wallabies that prefer rocky outcrops, and the tree-climbing tree-kangaroos of Australia's northern rainforests.

Rodents are eutherian mammals, not marsupials. As such, the young develop within the body for a longer period, connected to a placenta that feeds the developing foetus, removes waste and aids respiration, and they give birth to well-developed young. There are 52 native and seven introduced rodents in Australia, and a further 14 are now classed as extinct. They resemble the marsupial dasyurids in body shape and overall appearance, but differ mainly by having a single pair of chisel-shaped teeth (incisors), as opposed to the dasyurids' multiple rows of sharply pointed incisors.

Bats are the largest assemblage of mammals in Australia, with 77 species currently recognized, and a further three listed as extinct. The group comprises the large, often noisy fruit-bats (or flying-foxes), Ghost Bat, horseshoe-bats, leaf-nosed bats, sheath-tailed bats, free-tailed bats, bent-winged bats and the largest group, the vespertilionid bats. Bats have two teats, located near the armpits, and usually produce a single offspring each year.

Marine mammals include dugongs, seals, whales, dolphins and killer whales. This large group of about 60 mammals is aquatic, and all except the seals spend their lives in the water, eating, mating and giving birth there. Many are rare visitors to Australian waters, while others are seldom sighted, often only being seen when washed up on beaches, sick, injured or dead. They are specially adapted to their watery homes, with sleek, muscular bodies, and large flippers and tails. Their noses are located on top of the head to make breathing easier and safer. Seals come to shore to bask, sleep, mate and give birth, and appear somewhat cumbersome when on land, but they are truly at home in the water and superbly adapted for their semi-aquatic life.

Other mammals The remaining native Australian mammal species form small or individual family groups, including the Numbat, marsupial moles, Koala, Greater Bilby and Musky Rat-kangaroo. The Koala is famed throughout the world and is often mistakenly called a bear. It lives solely on a low-energy diet of eucalypt leaves and spends up to 20 hours a day sleeping. It is perhaps more sloth-like in habits than bear-like, but is nonetheless a marsupial that has a pouch for protecting the newborn young (joeys), which are born in the very early stages of development, and develop in the mother's pouch for about six months. Bears, on the other hand, are eutherian mammals that have a placenta to nourish the young as they develop within the body, and give birth to their young at a well-developed stage.

EXTINCTIONS AND INTRODUCTIONS

Extinction occurs when all members of a species die. To date, about 30 modern endemic Australian mammals have been classified as extinct since European settlement, with over 100 more being classed as Critically Endangered, Endangered or Vulnerable, most of which are rapidly declining in numbers or severely restricted in range. Many species occur today in isolated and fragmented areas of suitable habitat within their range, which leaves these subpopulations susceptible to disease and deformities from inbreeding.

Perhaps the most famous extinct Australian mammal is the Thylacine, or Tasmanian Tiger *Thylacinus cyanocephalus*, which was persecuted by farmers for eating their sheep and had a bounty of £1 per adult and 10/- (shillings) per young placed on it by the Tasmanian government in 1888 to encourage hunters to cull its numbers. The last known Thylacine died in Hobart Zoo on 7 September 1936, and despite numerous reported sightings, both in Tasmania and on mainland Australia, comprehensive field surveys over recent years using an extensive network of baited camera traps have been unable to confirm its continued existence.

The most recent Australian mammal to be declared extinct was the Bramble Cay Melomys *Melomys rubicola*, which inhabited a small, well-vegetated coral island (Bramble Cay) in the Torres Strait. After extensive surveys of the island, it was declared extinct in 2014.

Artist's illustration of the extinct Thylacine.

Australian Mammals, Past & Present

Introduced predators, primarily the Domestic Cat *Felis silvestris catus* and Red Fox *Vulpes vulpes*, have had an enormous impact on native mammals and other animals, with an estimated 400 vertebrate species being hunted as part of their collective diets, and the Cane Toad *Rhinella marina*, which has spread throughout much of northern and eastern Australia, has powerful toxins that kill almost every animal that eats it. Today, Australia has 32 introduced mammal species, which predate on native mammals, compete for food and shelter, or have negative impacts on the structure and composition of the natural environment that native mammals depend upon for survival. The number of feral cats in Australia alone has been conservatively estimated at about 4 million, with each cat killing up to an estimated 30 animals a day, and they have become established in every habitat throughout the continent.

Trapped feral cat.

Dingo near Fogg Dam in Australia's Northern Territory.

Australian Mammals, Past & Present

Another introduced mammal predator in Australia is the Dingo *Canis familiaris*, which lived in isolation in Australia after its presumed introduction about 3,500 years ago. Given the extent of its existence in Australia, it is now considered native. There are differing views on the taxonomy and nomenclature of the Dingo, with some proponents recommending that it be awarded species level classification as *Canis dingo*. We follow Jackson et al. (2017) in including Dingoes with other domestic dogs that are descended from the Gray Wolf *C. lupus*. Dingoes interbreed with other domestic dogs, although the majority of the population is still purebreed.

AUSTRALIA'S WHALING HISTORY

The Southern Right Whale, a regular visitor to the marine waters of southern Australia, earned its common name because it was considered the 'right' whale to hunt, a sad claim to fame that had both it and the Northern Right Whale close to extinction. The whales were deemed the 'right' whales to hunt as they formed large groups close to shore, so that they were easily accessible to 18th- and 19th-century whalers who used handheld harpoons and small boats, and the whales floated after being harpooned, making them easier to tow back to shore.

Whaling became a major primary industry, and whale products, including soaps, candles, lamp oil, perfumes, clothing and umbrellas, were sold locally and exported overseas. Numerous whaling stations were established along the Australian coastline at ports including Eden, Port Lincoln, Port Fairy, Portland and Albany.

Minke Whales on board a whaling ship in the Southern Ocean.

Australia's Vegetation

In the mid–late 19th-century, steam-driven boats and explosive harpoon guns revolutionized the whaling industry, and previously out-of-reach species, including Sperm Whales, Blue Whales and Humpback Whales, soon became targets for whaling fleets. Before 1900, when the whaling industry started to focus on Blue Whales, the species was regarded as abundant, with an estimated 350,000 population worldwide. The species was given worldwide protection in 1966, but by this time Blue Whale numbers had been reduced to about 1 per cent of the existing population size – today, the global population is thought to be about 10,000–25,000 individuals. Humpback Whales were considered to be close to extinction at around the same time, with an estimated 90 per cent of the population being killed, and only around 200-300 individuals left in Australian waters. Huge numbers of Sperm Whales were also hunted worldwide, and around 67 per cent of their estimated global population is believed to have been killed. Whaling ceased in NSW in 1962, and the rest of Australia followed by 1978.

Australia's Vegetation

Australia's landscape has evolved over the past 3,000 million years. It has been shaped by major geological change, resulting from continental drift, changing sea levels, long-term wind and water erosion, and the more recent volcanic activity of only a few thousand years ago. The climate has also become drier over the past 25 million years, leading to increasing soil aridity, and altering the distribution and composition of the native vegetation. It will continue to undergo change as the continent moves further north.

Tropical rainforest of northern Queensland.

AUSTRALIA'S VEGETATION

Tall open forest in Tasmania.

At localized levels, Australia is home to many thousands of vegetation types. The following broad groups are the ones most widely referred to throughout this guide and broader published material.

RAINFORESTS

Including tropical, subtropical, temperate and monsoon vine thickets. Found in the wetter climatic zones of Australia and typically characterized by dense foliage and a high diversity of plant species.

OPEN FORESTS

Open forests generally consist of forests with trees 10–30m in height. They are widespread in eastern (including Tasmania), northern and south-western Australia, and have a shrubby or grassy understorey. **Tall, open forests** typically have trees more than 30m in height (Australia's tallest species is the Mountain Ash *Eucalyptus regnans*, reputed to be the tallest hardwood tree in the world, with individual trees growing to more than 100m in height); they are found in wetter regions of the country. **Low, open forests** comprise smaller trees, with an average height of about 5–10m, and are generally found in areas that are cooler, drier and lower in nutrients, and have rocky slopes or are subject to regular flooding.

■ Australia's Vegetation ■

Open woodland.

Chenopod shrubland in South Australia.

AUSTRALIA'S VEGETATION

WOODLAND/OPEN WOODLAND

Woodland can contain a diverse assemblage of mixed tree species and many are restricted in total range. Other woodland is dominated by a single plant genus, including eucalypts, acacias, cypress, casuarinas and melaleucas. Open woodland has a wider spacing between trees, allowing neighbouring grassland and shrubland to invade, and forming a valuable mosaic of these different vegetation communities.

Montane heath in Tasmania.

SHRUBLAND

Typified by multi-stemmed shrubs, either monotypic or a broad range of shrub species. The dominant shrubland type in Australia is acacia, including mulga and gidgee, with other shrubland having a mix of grevilleas, samphires, saltbush, chenopods, banksias and emu bushes.

HEATHLAND

Typically a mixture of species, many with a mature height of 1m or less, with dense canopies. Associated with low-nutrient soils, including coastal montane, laterite and sandy soils, or areas subject to erosion or waterlogging.

Hummock grassland near Uluru, Northern Territory.

Australia's Vegetation

Mangroves.

Paperbark swamp.

Australia's Vegetation

Open rocky area.

GRASSLAND

Generally dominated by herbaceous (non-woody) species and occurring in a range of areas. Two main communities are found extensively in Australia, these being tussock and hummock grassland. Tussock grassland is characterized by a broad range of perennial grasses growing in tufts, including Mitchell Grass and Blue Grass, while hummock grassland is dominated by Spinifex *Triodia* and *Plechrachne* spp., and is typical of the arid lands of Australia.

Marine habitat.

Evergreen perennials form mounds up to 1m in height, with areas of open, exposed and usually bare soil in between hummocks. Soils are typically sandy or rocky (skeletal), and either hilly or flat.

MANGROVES

Found in the intertidal zone in coastal areas that are protected from high waves. Mangroves tend to form tall, closed forests in the north, and low, open forests or shrubland in the south.

Australia's Vegetation

INLAND WATERWAYS

A mixture of fresh and brackish aquatic areas, including rivers, creeks, billabongs, lakes, swamps and marshes.

BARE GROUND

Areas that are largely lacking in vegetation, or containing some pioneer plant species. They can be in the form of exposed rock, coastal sands and dunes, desert sands and claypans. The soils have a low nutrient content and are prone to erosion.

MARINE

Open water in oceans and estuaries bounded by the Australian continental shelf. Australia's oceans and seas are the Pacific, Southern and Indian Oceans, and Timor, Tasman and Coral Seas. Australia's marine zone extends to the edge of its continental shelf, occupying a total marine area of about 8 million square kilometres, including remote offshore territories, and a further 2 million square kilometres in the Australian Antarctic Territory.

URBAN

These are diverse habitats and include remnants of bushland areas (as described above), but also contain wasteland, dwellings and gardens.

Eastern Grey Kangaroos in a NSW garden.

◾ Using This Book ◾

Using This Book

Due to variations in sizes within subspecies and geographically isolated populations, the sizes shown in the species accounts are approximates of the average maximum sizes for full-grown adults, unless otherwise indicated. For most species, this is given as Total Length (TL), which includes the length of the tail, where relevant. Many similar species are distinguished partly by the ratio of the length of the tail to the animal's head-body length, so the length of the tail is also shown. Similarly, the Forearm (FA) length is used to externally identify certain similar bat species, and this measurement has been included within these species accounts.

Alternative names of species, where relevant, are provided in parentheses under the main species headings.

The species accounts and checklist at the rear of the book follow the naming protocols and species sequence within the major taxonomic reference used for this book (Jackson & Groves, 2015).

KEY FEATURES AND MEASUREMENTS

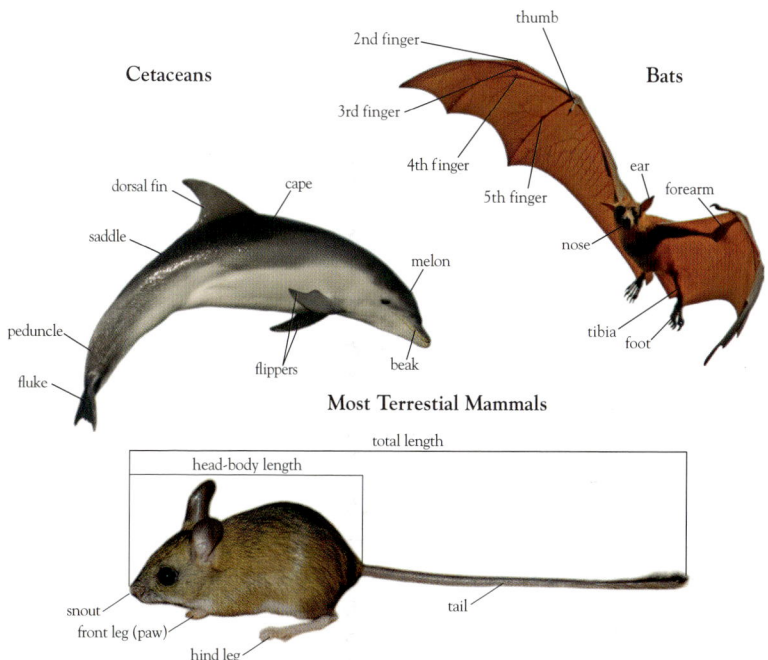

Glossary

DISTRIBUTION KEY

The following abbreviations for localities with the Australian geographic zone have been used in this book:

ACT	Australian Capital Territory	**NSW**	New South Wales
NT	Northern Territory	**Qld**	Queensland
SA	South Australia	**Tas**	Tasmania
Vic	Victoria	**WA**	Western Australia

Glossary

alpha (male) Dominant (sexually) male within a group.

anterior Front of the body.

aquatic Living in fresh or salt water.

arboreal Living in or climbing trees for food or shelter.

attenuate Narrow gradually.

baleen Comb-like plates attached to upper jaw of some whale species, used to filter krill, plankton and small organisms from the water.

beak Elongated front portion of head or jaw of animals.

benthic Lowest section of ocean or lake.

blowhole Opening (nostrils) of dolphin or whale at top of head, used for breathing.

bow-riding Riding the waves formed by a water vessel.

breach To leave the water via swimming (generally applicable to cetaceans).

callosities Calcified skin patches found on cetaceans that are unique to the individual and used for identification.

camps (colonies) Large roosting groups (particularly of bats).

canines Pointed teeth between incisors and premolars.

cape Cetacean's back or dark section over shoulders.

carnivorous Feeding only on animal matter.

carrion Flesh of dead animals.

cathemeral Irregularly active at any time of night or day, depending on circumstances (*see also* diurnal and nocturnal).

crepuscular Active at dawn and dusk.

crustacean Invertebrate group with hard exoskeleton (such as shrimp and lobsters).

diagnostic Characteristic that defines identification of a species.

digit Toe or finger.

distal Section of limb or attachment furthest from body.

diurnal Active during the day.

dorsal Upper surface, or back.

echolocation Using sound emitted, and analysing returning echo, to determine location of prey or objects.

endemic Found only in a certain area.

epiphytes Plants (non-parasitic) growing on other plants.

exudate Substance secreted by a plant or insect (such as sap or honeydew).

Glossary

feral Introduced animal that has become established in the wild.
frugivorous Feeding mostly or entirely on fruits.
genus (genera) Taxonomic group above species and below family.
gregarious Living within a group or community.
herbivorous Feeding only on plant matter.
hibernation Period of inactivity (normally during winter).
honeydew Sugar-rich liquid.
incisors Chisel-shaped front teeth used for biting or gnawing.
insectivorous Feeding solely on insects.
invertebrate Animal that lacks a backbone.
krill Planktonic shrimp (the main food supply of baleen whales).
larva Newly hatched, wingless stage of an insect.
membrane Thin layer of tissue.
mob Small social group (specifically of macropods).
molars Rear teeth use to chew or grind.
nectarivorous Feeding purely on nectar.
nocturnal Active at night.
nominate (subspecies) Having the same epithet (name) as the species.
noseleaf Naked plates around nostrils of bats used in echolocation.
omnivorous Feeding on both animal and plant matter.
pelagic Living in open ocean.
pelt Fur or skin of an animal.
plankton Group of microscopic organisms that float in water current.
pod Small group of cetaceans.
posterior Rear of body.
prehensile (tail) Adapted to grasp or attach to objects.
riparian Found on banks of rivers, streams, creeks and similar places.
scat Mammal droppings.
species Basic unit of taxonomic classification.
spy-hopping Cetaceans' method of viewing above the water by rising vertically.
subspecies Level of taxonomic division below species.
synonymous Considered to be the same species.
taxonomy Classification of living things based on characteristics.
teat Nipple from mammary gland.
terrestrial Living on or spending time on the ground.
territory Area that an individual or group occupies and protects.
torpor State of inactivity or lethargy, similar to hibernation.
ventral Undersurface or belly of an animal.
vertebral Along line of spine (vertebrae).
vertebrates Animals that have a backbone.
vestigial Remnant of an appendage that has lost its original use through evolution.
weaning Movement from feeding on mother's milk to other food.

Ornithorhynchidae, Platypus

Platypus ■ *Ornithorhynchus anatinus* TL 390–600mm

DESCRIPTION Streamlined body with a broad flat tail covered with dense, dark greyish to reddish brown waterproof fur above, with flat, paddle-shaped tail and broad, greyish-brown 'duck-like' bill that extends over forehead and chin. Also has short limbs with webbed feet. Underparts paler silver greyish or reddish-brown. Males possess a poisonous horny spur on the ankles of each hindleg, which connect to a venomous gland further up the leg. **DISTRIBUTION** Distributed through eastern and south-eastern Australia. **HABITAT AND HABITS** Found in regularly flowing river systems and associated billabongs, where it hunts along the bottom of the waterway for a variety of large aquatic invertebrates. Bill has electrosensors and mechanoreceptors to detect electrical pulses and movement of prey. Nesting chamber is at the end of a long burrow with an entrance just above the water level. Up to 3 (usually 2) soft-shelled eggs are incubated for about 12 days before they hatch.

Tachyglossidae, Echidna

Short-beaked Echidna ■ *Tachyglossus aculeatus* TL 300–450mm

DESCRIPTION Pale brown to blackish fur above and below, which varies in length depending on subspecies, and numerous protective spines on upper body. Snout long and cylindrical, covered with sensitive skin, with nostrils and small mouth at tip. **DISTRIBUTION** Australia-wide, including major islands. **HABITAT AND HABITS** Occurs in variety of habitats, from wet forests to deserts. Mostly crepuscular and nocturnal, although active by day in cooler areas. Feeds on ants, termites and other invertebrates, which are collected using its long, sticky tongue. Single soft-shelled egg hatches after about 10 days, and young 'puggle' is carried in mother's pouch for up to 2 months. If threatened, curls into loose ball or digs itself into the ground, leaving only the sharp spines exposed.

Tasmanian subspecies

DASYURIDAE, DASYURIDS

Brush-tailed Mulgara
■ *Dasycercus blythi* TL 200–235mm, including tail 75–100mm
(Murtja)

DESCRIPTION Generally pale yellowish-brown above, and tail with reddish-brown base and black brushy tip. Underparts greyish-white. Female has pouch. Limbs short with 5 toes on each foot. Ears short and rounded, and snout pointed. **DISTRIBUTION** Arid sandy inland areas, almost reaching coast in central WA, through southern NT and into far western Qld. Exact range poorly known, due to previous grouping with Crest-tailed Mulgara (see below). **HABITAT AND HABITS** Found in hummock grassland, and adjacent woodland and open sandy plains. Mostly sleeps in deep, complex burrow, and forages mainly at night for large invertebrates and small vertebrates.

Crest-tailed Mulgara
■ *Dasycercus cristicauda* TL 250–355mm, including tail 100–125mm
(Ampurta)

DESCRIPTION Generally pale yellowish-brown above. Tail has reddish-brown base and latter two-thirds black, with hairy dorsal crest towards tip. Underparts greyish-white. Female has pouch. Limbs short with 5 toes on each foot. Ears short and rounded, and snout pointed. **DISTRIBUTION** Arid deserts of south-eastern NT, north-eastern SA and far south-western Qld. **HABITAT AND HABITS** Found in sparsely vegetated fringes of salt lakes and sand dunes. Hunts around dunes, mostly at night, for range of invertebrates and small vertebrates, supplemented with some fruits and seeds.

▪ Dasyurids ▪

Kaluta ▪ *Dasykaluta rosamondae* TL 150–180mm, including tail 55–70mm

DESCRIPTION Stocky dasyurid with thickened, tapering tail. Uniformly reddish-brown (including tail), with long, shaggy fur, short, rounded ears and short, pointed snout.
DISTRIBUTION Arid central western WA, including Little Sandy Desert and Pilbara.

HABITAT AND HABITS Found in hummock grassland. Forages at night in warmer months and by day in winter, for insects and other invertebrates and some small reptiles, sleeping at other times in underground burrows. Males die shortly after mating. **Notes** Undergoes nocturnal torpor during cold winter nights.

Kowari ▪ *Dasyuroides byrnei* TL 245–320mm, including tail 110–140mm

DESCRIPTION Stocky. Pale grey-brown, with white ring around each large eye, pointed snout and thin, mostly naked, pinkish ears. Tail has white base and distinctive black, 'brush-like' distal half. Limbs white, with 5 toes on each hindfoot. **DISTRIBUTION** Confined to Sturt's Stony Desert in south-western Qld and north-eastern SA. **HABITAT**

AND HABITS Inhabits arid gibber plains with sparse vegetation. Shelters by day in simple burrow or network of burrows. Hunts on the ground for small vertebrates and large invertebrates. If threatened, hisses and flicks tail, and can climb or leap large distances to avoid danger. Population size fluctuates greatly in response to rainfall and associated abundance of prey.

▪ DASYURIDS ▪

Western Quoll
■ *Dasyurus geoffroyi* TL 470–750mm, including tail 210–350mm (Chuditch; Idnya)

DESCRIPTION Brown to reddish-brown above, with numerous white spots on back, sides and flanks. Tail long and hairy, being mostly blackish, especially towards tip, and occasionally with some white spots on base. Underparts whitish. **DISTRIBUTION** Restricted to timbered areas and wheatbelt of far south-western WA. **HABITAT AND HABITS** Occurs in wet and dry areas in sclerophyll forests and woodland with dense groundcover, and adjacent habitats. Shelters by day in burrows, hollow logs, rock crevices or tree hollows. Hunts, mainly on the ground, for wide range of large invertebrates and small vertebrates, supplemented with plant matter and carrion.

Northern Quoll
■ *Dasyurus hallucatus* TL 250–620mm, including tail 125–310mm (Satanellus; Njanmak)

DESCRIPTION Brownish above, with numerous large white spots, and off white below. Tail unspotted, although some spots may occur on base, with darker brown to blackish tip. **DISTRIBUTION** Isolated populations across tropical northern Australia (including several offshore islands), in Pilbara and Kimberleys WA, northern NT and eastern Qld. **HABITAT AND HABITS** Found in variety of habitats, including open forests, woodland, scrubland and grassland, preferring rocky areas. Shelters by day in hollow logs, termite mounds, rocky crevices or tree hollows. Omnivorous, feeding at night on small vertebrates, including mammals, reptiles and amphibians, invertebrates and some fruits.

■ Dasyurids ■

Spotted-tailed Quoll
■ *Dasyurus maculatus* TL 690–1,310mm, including tail 340–550mm
(Tiger Quoll; Yarri)

DESCRIPTION Large quoll with varying sized white spots on back, sides, legs (lower legs) and tail. Remaining fur rufous to dark brown above, and paler yellowish-grey below. Five toes on each foot. **DISTRIBUTION** Atherton Tablelands of north-east Qld and south-eastern Australia, from south-eastern Qld, through eastern NSW and ACT, to western Vic and Tas. Isolated records from central NSW and central Vic. **HABITAT AND HABITS** Found in wet tropical forests (in north), closed and open forests, woodland and coastal heathland. Forages on the ground or in trees for small to medium mammals, reptiles, birds and invertebrates, but also feeds on carrion. Mostly nocturnal, and shelters at other times in hollow logs, rocky crevices, tree hollows and burrows.

Eastern Quoll
■ *Dasyurus viverrinus* TL 450–730mm, including tail 170–280mm
(Luaner)

DESCRIPTION Two colour morphs: pale brown to greyish, and blackish, both with white spots on back and sides, and unspotted tail. Underparts off white. Ears larger and snout more pointed than Spotted-tailed Quoll's (see above), and 4 toes on each hindfoot. **DISTRIBUTION** Confined to Tas. **HABITAT AND HABITS** Occupies wet and dry forests, heathland and dry grassland. Mostly nocturnal, resting by day in nest in hollow log, under rocks or in simple underground burrow. Forages on the ground for large insects, small vertebrates, fruits and some carrion.

Dark morph

Pale morph

■ DASYURIDS ■

Sandstone Pseudantechinus
■ *Pseudantechinus bilarni* TL 140–225mm, including tail 80–115mm

DESCRIPTION Greyish-brown above, more reddish-brown behind ears and with blackish patch on forehead. Tail moderately long and thinly tapering, around same length as head-body. Underparts greyish-white. Female lacks pouch. **DISTRIBUTION** Tropical northern NT, including central and western Arnhem Land, Marchinbar Island and south-west hinterland of Gulf of Carpentaria. Overlaps in east of range with similar Carpentarian Pseudantechinus (see p. 26). **HABITAT AND HABITS** Nocturnal, agile climber, inhabiting rocky sandstone formations with large boulders and abundant crevices, with varying vegetation cover. Forages for invertebrates and small vertebrates.

Fat-tailed Pseudantechinus
■ *Pseudantechinus macdonnellensis* TL 170–190mm, including tail 75–85mm

DESCRIPTION Mostly greyish-brown above with reddish-orange behind ears, and underparts greyish-white. Tail normally swollen at base and slightly shorter than head-body length. Head has pointed snout and large eyes. **DISTRIBUTION** Arid central western inland, including central WA, northern SA and southern NT. **HABITAT AND HABITS** Occurs on lightly vegetated rocky slopes and in sandy plains. Sleeps by day in nest in rocky crevices or termite mounds, and forages mainly at night for small invertebrates. May also bask during the day. Female carries young in moderately developed pouch. Stores fat in base of tail during times of plentiful food.

Dasyurids

Carpentarian Pseudantechinus
Pseudantechinus mimulus TL 130–165mm, including tail 60–75mm

DESCRIPTION Brown above, streaked paler and darker brown, with reddish patch behind each ear, and pale grey to whitish below. Tail shorter than head-body length, generally swollen at base and strongly tapering. **DISTRIBUTION** Restricted to Sir Edward Pellew Group of Islands in Gulf of Carpentaria NT, and ranges in Gulf hinterland, NT and Qld. **HABITAT AND HABITS** Found in rugged sandstone (mainly) and limestone formations. Forages at night for invertebrates and small vertebrates, and likely to shelter in rocky crevices by day.

Ningbing Pseudantechinus
Pseudantechinus ningbing TL 150–190mm, including tail 75–95mm

DESCRIPTION Generally pale greyish-brown above with pinkish-cinnamon patches behind ears, and whitish-grey to buff below. Tail moderately well furred, usually swollen at base, sharply tapering and similar in length to head-body. Head has large ears and dark line running from tip of pointed snout to back of neck. **DISTRIBUTION** Kimberley region of northern WA, east to Gregory National Park NT. **HABITAT AND HABITS** Occurs in variety of habitats, including low eucalypt woodland with grassy spinifex understorey, on rocky sandstone and limestone outcrops. Forages at night for insects.

Dasyurids

Woolley's Pseudantechinus
Pseudantechinus woolleyae TL 175–195mm, including tail 75–85mm (Woolley's False Antechinus)

DESCRIPTION Rich brown above, with reddish-brown patch behind ears, and pale pinkish-brown below. Tail usually swollen at base and about 75 per cent of head-body length. Snout sharply pointed. Larger and less reddish than very similar **Tan Pseudantechinus** *P. roryi*. **DISTRIBUTION** Pilbara region and arid central west of WA. **HABITAT AND HABITS** Occurs in various habitats, including hummock grassland, acacia scrubland and open plains, where it favours rocky slopes, and forages at night.

Tasmanian Devil
Sarcophilus harrisii TL 815–910mm, including tail 245–260mm

DESCRIPTION Unmistakable. Predominantly black, occasionally with reddish wash, and white patches on chest, and occasionally on shoulders and rump. Head wide (wider in males than females), with large, powerful jaws. Emits loud, guttural vocalizations.

DISTRIBUTION Restricted to Tas, including Robbins and Maria (introduced in 2012) Islands. **HABITAT AND HABITS** Found in most habitats, including wet and dry sclerophyll forests, woodland, scrubland and grassland. Nocturnal, foraging on the ground for medium-sized mammals, birds, large invertebrates and carrion, with all parts of the animal (even bones) digested. Sleeps by day in hollow logs, rocky caves, dense vegetation or den. Population decimated by Devil Facial Tumor Disease (DFTD), an infectious cancer that is spread by bites.

▪ DASYURIDS ▪

Rusty Antechinus
▪ *Antechinus adustus* TL 180–220mm, including tail 90–105mm

DESCRIPTION Fur moderately long, and uniformly dark brown above, with reddish-brown tips to hairs on head and back, and more cinnamon below, extending onto flanks. Tail has blackish tip. No reddish-brown behind ears or on cheeks, and lacks pale eye-ring.

DISTRIBUTION Wet tropics of north-eastern Qld, between Bluewater Range and Mount Windsor Tableland, generally above 600m. **HABITAT AND HABITS** Occurs in fringes of undisturbed dense upland vine forest. Nocturnal and diurnal, foraging both on the ground (usually around rotting logs) and in trees for invertebrates and small vertebrates. Sleeps at night in tree hollows and tree ferns. All males die shortly after mating.

Agile Antechinus
▪ *Antechinus agilis* TL 150–220mm, including tail 75–110mm

DESCRIPTION Pale brownish-grey above, with paler ring around eye, and greyish below. Head has pointed snout and large, rounded, pinkish-grey ears. Tail thin and lightly furred.

DISTRIBUTION Near coast and ranges of south-eastern mainland, from southern NSW and ACT to western Vic. **HABITAT AND HABITS** Found in forests, woodland and heathland. Forages in trees and on the ground, in leaf litter and around fallen logs, for insects and small vertebrates, and shelters communally at other times in tree hollows. Males die shortly after breeding, but some females survive to breed a second year.

DASYURIDS

Black-tailed Antechinus
■ *Antechinus arktos* TL 200–276mm, including tail 94–131mm

DESCRIPTION Large, with shaggy fur. Generally brownish with longer fuscous to orange-brown guard hairs. More greyish on head, with orange-brown fur on cheeks and upper and lower eyelid. More orange-brown on rump, with black tail, and greyish-black on tops of hindfeet. Underparts dark grey to olive-grey. **DISTRIBUTION** South-eastern Qld and north-eastern NSW. **HABITAT AND HABITS** Occurs in rainforests and wet sclerophyll forests above 780m. Forages in vegetation and around buttress roots of trees, seemingly around creeks, for invertebrates. Males probably die shortly after mating.

Fawn Antechinus
■ *Antechinus bellus* TL 205–275mm, including tail 95–125mm

DESCRIPTION Greyish above, occasionally washed with pinkish-brown, and cream to greyish below. More white on chin and feet, and males may be stained with yellow around gland on chest. No reddish-brown behind ears or on cheeks. **DISTRIBUTION** Tropical northern NT and adjacent Melville Island. **HABITAT AND HABITS** Found in mature lowland open forests and woodland with dense shrubby understorey. Crepuscular and nocturnal, sleeping by day in tree hollows or hollow logs, and emerging to forage on the ground or in trees for invertebrates and small vertebrates, including geckoes. Males die shortly after mating, and females survive to breed for second and (rarely) third season. Absent or numbers greatly reduced in areas affected by frequent or intense fires.

DASYURIDS

Yellow-footed Antechinus
■ *Antechinus flavipes* TL 150–320mm, including tail 65–150mm
(Mardo)

DESCRIPTION Very colourful compared to other Antechinus spp. Distinctly grey head with a pointed yellowish snout and orange to yellowish-brown sides, with yellowish belly, rump and feet. Feet may be whitish to yellow depending on geographic location. Perimeter around eyes presents a white to grey eyering and a distinctly black tip to tail. Variable across Australia; in north Qld they are large and reddish-orange, in south-west WA they are a lighter brown with a whitish colour on the belly and feet. **DISTRIBUTION** This is the most widespread Antechinus sp. and occurs from north-east Qld (Cooktown to near Ingham) then in a broad band from Airlie Beach down the east coast to Vic and SA ,and west to Central Qld, far west NSW and Vic. In addition, there is also a population in south-west WA. **HABITAT AND HABITS** Occurs in tropical vine forests, rainforests, dry open forests, woodland, swampland and mulga. Sleeps during most of the day in a crevice in the ground or in a burrow. Nests are made of leaves in hollow trees, rocks and on the ground in hollows. Forages on the ground, shrubs and trees for insects, small invertebrates, vertebrates, flowers and nectar. Also found to devour small birds and mice. Very active and more diurnal than other antechinus species.

Mainland Dusky Antechinus
■ *Antechinus mimetes* TL 161–287mm, including tail 76–122mm

DESCRIPTION Large and dark. Predominantly brownish, paler on belly. Broad head,

pointed snout and tail shorter than head-body length. **DISTRIBUTION** Two subspecies, *A. m. mimetes* on eastern Great Dividing Range, south-eastern NSW, ACT and southern Vic, and *A. m. insulanus* in Grampians Range, western Vic. **HABITAT AND HABITS** Within Vic, found in wetter areas, including rainforests, woodland and alpine heaths, and similar habitats in NSW, but also in coastal sand dunes and swamps, where it feeds on invertebrates. Males die shortly after mating.

◾ Dasyurids ◾

Swamp Antechinus
◾ *Antechinus minimus* TL 163–250mm, including tail 62–98mm

DESCRIPTION Metallic greyish on head and shoulders, becoming more yellowish-brown on back, sides and thickset rump, and darker brown on short tail. Underparts pale brown to yellowish, tinged with grey. **DISTRIBUTION** Two subspecies recognized, *A. m. minimus* in Tas (including islands of Bass Strait) and *A. m. maritimus* in coastal Vic and far south-eastern SA. **HABITAT AND HABITS** Found in rainforests and open habitats, including heathland, tussock grassland, swamps and shrubland, where it shows a preference for wetter areas. Sleeps in nest placed beneath dense leaf litter or in short underground burrow. Forages mainly at night for range of insects. Males die shortly after mating, but some females breed a second time. Population sizes appear to increase during wetter periods.

Buff-footed Antechinus
◾ *Antechinus mysticus* TL 210mm, including tail 110mm

DESCRIPTION Greyish-brown on head and shoulders, changing to more yellowish-brown on lower back and rump, and tail brownish with darker tip. Similar to Yellow-footed Antechinus (see p. 30) but duller. **DISTRIBUTION** Eastern Qld from near NSW border to around Mackay. **HABITAT AND HABITS** Lives in rainforests and open, grassy woodland. Forages at night for range of invertebrates.

DASYURIDS

Brown Antechinus

■ *Antechinus stuartii* TL 140–250mm, including tail 65–110mm
(Berruth)

DESCRIPTION Greyish-brown above washed with reddish-brown and paler brown below. Long pointed head, which tends to be broad across the upper head, and four pairs of incisors. Nose is pinkish. Tail is thin, sparsely haired and same length or slightly shorter than body. Eyes are bulging, dark and quite prominent for its size. Ears are thin with a notch in the margin. In wet habitat females have six nipples but can have ten in the drier and higher altitude habitats. **DISTRIBUTION** Eastern mainland from east of

Warwick in Qld across to Byron Bay in northern NSW and south to Batemans Bay, NSW, and ACT, and as far west as Bathurst NSW. Overlaps in range with Agile and Mainland Dusky Antechinuses (see pp. 28 and 30), and **Subtropical Antequinus** *A. subtropicus*.
HABITAT AND HABITS Prefers wet sclerophyll forests with dense ground cover and fallen timber, which they can use as cover and for nesting. Generally ground dwelling but also arboreal in drier habitats where the ground cover is sparse, or where competition for food is greater. Carnivorous and mainly nocturnal. Preys upon insects, spiders, centipedes, small reptiles and occasionally frogs. Some individuals may become more active during the day in the winter period when the food resource is limited.

Tasmanian Dusky Antechinus

■ *Antechinus swainsonii* TL 180–271mm, including tail 77–110mm

DESCRIPTION Generally greyish-brown, darker on back and paler on sides, with brownish wash on rump and greyish-white underparts. Head has very slender snout, and tail significantly shorter than head-body. **DISTRIBUTION** Throughout most of Tas, except Tasman Peninsula. **HABITAT AND HABITS** Found in range of forested habitats, usually with dense understorey. Forages at night for invertebrates, small vertebrates and fruits, found by fossicking in the ground litter and soft soils. Males die shortly after mating.

◾ Dasyurids ◾

Brush-tailed Phascogale
◾ *Phascogale tapoatafa* TL 310–495mm, including tail 160–235mm
(Wambenga; Tapoa Tafa)

DESCRIPTION Grey above, streaked with silvery-white, and white to creamish below, with long, bushy black bottlebrush tail. Head with a pointed snout, large naked ears and large protruding eyes. Long sharp claws and hind feet have a striated footpad. **DISTRIBUTION** Patchily distributed in south-western and the Kimberley region WA, north-eastern and south-eastern Qld, eastern NSW and central and south-western Vic. **HABITAT AND HABITS** Prefers dry sclerophyll open forests with herbs, shrubs, grass and leaf litter. Also occurs to a lesser degree in rainforests, wet sclerophyll forests, heath and swamps. Generally most numerous in moist gullies, where it shelters in tree hollows and emerges at night to forage almost exclusively within trees for invertebrates concealed under loose bark, supplemented with nectar. This species is the most arboreal of the dasyurids and is an agile climber, capable of large acrobatic leaps between branches. When alarmed, it taps its front feet. All males die after breeding.

Gile's Planigale ◾ *Planigale gilesi* TL 115–150mm, including tail 55–70mm

DESCRIPTION Grey above, washed with yellowish-brown, and cream to pale brown below, paler on throat and sides of neck. Body flattened, with short legs, pointed snout and small, rounded ears. **DISTRIBUTION** Arid and semi-arid central eastern mainland Australia, from south-eastern NT and south-western Qld, through northern and western NSW, to north-eastern SA. **HABITAT AND HABITS** Found in floodplains and dry channels with dense vegetation and heavy cracking clay soils. Sleeps by day in soil crevices, and forages on the ground for insects. May also hunt by day or bask in the sun during cold weather. Shares range with Narrow-nosed Planigale (see p. 35), but is more crepuscular.

DASYURIDS

Long-tailed Planigale
■ *Planigale ingrami* TL 101–123mm, including tail 44–60mm

DESCRIPTION Grey-brown, with long, thin tail, larger hindfeet than forefeet, and head very flattened, with pointed snout. **DISTRIBUTION** Northern Australia, from Kimberley region WA, through northern NT, to central western and eastern Qld. Absent from Cape York Peninsula and coastal Qld. **HABITAT AND HABITS** Occurs in tropical grassland, black soil plains and riverine systems. Sleeps by day in natural crevices, and emerges at night to forage for insects. May also hunt and bask in the sun on cold mornings.

Common Planigale
■ *Planigale maculata* TL 130–195mm, including tail 60–95mm

DESCRIPTION Pale grey to cinnamon-brown above, streaked with silvery-grey, and paler below. Head flattened and pointed, with large, rounded ears, and tail shorter than head-body length. Hindfeet broad and toes pale. **DISTRIBUTION** Northern and eastern mainland Australia. **HABITAT AND HABITS** Found in rainforests, wet sclerophyll forests, woodland, marshes and moist grassland. Sleeps by day in nest placed in natural crevice, such as under a log or rock, and forages for invertebrates and small vertebrates. Capable of eating prey almost as large as itself.

Dasyurids

Narrow-nosed Planigale
■ *Planigale tenuirostris* TL 100–140mm, including tail 50–65mm

DESCRIPTION Reddish-brown above and whitish below, with tail similar in length to head-body. Head and body flattened, snout pointed and toes yellowish-brown. Female develops temporary pouch during breeding season. DISTRIBUTION Arid and semi-arid central eastern inland, including south-eastern NT, north-eastern SA, western NSW and south-western Qld. HABITAT AND HABITS Found in floodplains and dry channels with dense vegetation. Sleeps by day in small burrow, and emerges at night to forage for insects. May also forage by day or bask in the sun during cold weather.

Kultarr
■ *Antechinomys laniger* TL 170–250mm, including tail 100–150mm (Wuhl-wuhl; Pitchi-pitchi)

DESCRIPTION Pale brownish-grey to yellowish-brown above and white below. Head darker on crown, with large ears and darker eye-ring. Limbs long, and tail has distinctive tufted blackish tip. DISTRIBUTION Arid to semi-arid inland areas from western WA, through SA, southern NT, south-western Qld and north-western NSW. HABITAT AND HABITS Occurs in sparsely vegetated hummock grassland, scrubland, gibber plains and deserts. Sleeps by day in hollow logs, soil cracks or burrows, or in bases of spinifex tussocks. Feeds at night, in clearings, on insects and other arthropods. Moves with bounding gait, pushing with long hindlegs and landing on forelegs.

◾ DASYURIDS ◾

Wongai Ningaui ◾ *Ningaui ridei* TL 116–145mm, including tail 60–70mm

DESCRIPTION Yellowish-grey to greyish-brown above, with longer black guard hairs, and ginger wash around bases of ears and on cheeks. Underparts whitish. Ears short and rounded. Tail thin, furred and similar in length to head-body. **DISTRIBUTION** Arid inland of south-western NT, north-western SA and central WA. **HABITAT AND HABITS** Favours arid to semi-arid spinifex grassland. Shelters by day in spinifex tussock or low shrub, emerging to feed at night on insects. Gains most of its moisture from food.

Pilbara Ningaui

◾ *Ningaui timealeyi* TL 105–140mm, including tail 60–80mm

DESCRIPTION Overall grizzled reddish-brown to grey-brown with rufous-washed face, ears and flanks. Lacks the reddish-brown fur around the head seen in Wongai Ningaui (above) of central Australia. Underside of hindfoot: two pads closest to heel, further back

than the pad at base of toe 1, are distinct, > 1mm long, and much longer than wide; foot usually short, <11.5mm. Females have six teats. **DISTRIBUTION** Pilbara and Gascoyne region WA. **HABITAT AND HABITS** Hummock grasslands with acacia or mallee scrub, where it favours drainage lines. Hunts at night for insects, and shelters at other times in hollow logs or under spinifex clumps. Partly arboreal. Breeding dependant on rainfall, and usually commences around September. Female has a shallow recess on her belly (no pouch), and the 4–6 young are weaned at around 13 weeks.

Dasyurids

Southern Ningaui
Ningaui yvonnae TL 105–145mm, including tail 55–70mm

DESCRIPTION Fur shaggy and somewhat spikey. Uniformly olive-grey above, washed dark reddish-brown. More greyish on head and with yellowish-brown on cheeks. Pale greyish below, becoming whiter on chin. Tail similar in length to head-body, thin and moderately furred. **DISTRIBUTION** Arid inland areas of southern mainland, from western WA, through southern SA to central NSW. **HABITAT AND HABITS** Occurs in spinifex grassland, and adjacent mallee shrubland and heathland, on sandy plains and dunes. Nocturnal and mostly terrestrial. Forages at night, mostly solitarily, for invertebrates and small vertebrates up to its own size. Shelters by day under dense vegetation, in hollow logs or shallow burrows. Enters torpor during colder months, or when food is scarce.

Kakadu Dunnart
Sminthopsis bindi TL 110–190mm, including tail 60–105mm

DESCRIPTION Pale brownish-grey washed with yellowish-orange, and more brownish on cheeks, with thin darker eye-ring, and occasionally obscure darker longitudinal stripe on forehead. Underparts and feet white. Tail thin and sparsely furred, and longer than head-body length (about 120 per cent). **DISTRIBUTION** Northern NT, including Wongalara Wildlife Sanctuary and Kakadu National Park. **HABITAT AND HABITS** Occurs in eucalypt woodland in sandstone escarpments, tending to prefer areas with gravel cover on the ground. Forages terrestrially at night for range of invertebrates, and shelters by day in small burrow.

▪ Dasyurids ▪

Butler's Dunnart
▪ *Sminthopsis butleri* TL 147–178mm, including tail 72–90mm
(Munjol)

DESCRIPTION Fur fine and soft, brownish-grey above, paler and more brownish on cheeks and lower sides, and white below. Tail similar in length to head-body, thin and sparsely furred. **DISTRIBUTION** Originally discovered in northern Kimberley WA, but

now thought to only inhabit Bathurst and Melville Islands NT. Previously thought to also occur on Cape York Peninsula, but specimens collected there since discovered to be **Chestnut Dunnart** *S. archeri*. **HABITAT AND HABITS** Found in tall eucalypt forests with dense grassy understorey. Hunts on the ground at night for insects and other arthropods.

Fat-tailed Dunnart
▪ *Sminthopsis crassicaudata* TL 100–160mm, including tail 40–70mm

DESCRIPTION Generally pale brown above and whitish below and on feet. Head has pointed snout, dark patch along sides of snout and around eyes, and sometimes white marks near bases of long, rounded ears. Tail greyish, and thickened towards middle.
DISTRIBUTION Southern and inland of eastern Australia, from coastal WA, through SA

and southern NT, western Vic, and inland of NSW and south-western Qld. **HABITAT AND HABITS** Occurs in various habitats, from wet to dry woodland, shrubland and grassland with cracking clay soils, to arid sandy deserts. Hunts at night for insects and other arthropods, and sleeps by day in grassy nest in soil cracks. Stores fat in tail for use during times of decreased food availability.

▪ Dasyurids ▪

Little Long-tailed Dunnart
▪ *Sminthopsis dolichura* TL 147–208mm, including tail 84–109mm

DESCRIPTION Pale to dark greyish above, extending to top of tail, and white below. Head paler, with long, naked ears, pale reddish-brown patches behind ears and on cheeks, and black eye-ring. Tail thin, and proportionally longer in males than females.
DISTRIBUTION Two geographically distinct populations in arid and semi-arid southern Australia, including southern inland WA (absent along southern coast and far east), and coastal southern SA. **HABITAT AND HABITS** Found in woodland, shrubland, heathland and grassland. Nocturnal, sleeping by day in nest of dry grass placed in hollow log or tussock, or in abandoned burrow, and foraging on the ground at night for insects and other invertebrates. Numbers tend to be greater in the 3–4 years after fires.

Julia Creek Dunnart
▪ *Sminthopsis douglasi* TL 210–265mm, including tail 110–130mm

DESCRIPTION Brown above, streaked with grey, with darker longitudinal stripe on forehead and forwards of eye, and reddish cheeks. Tail tapering, with blackish tip, and similar in length to head-body. Underparts whitish, tinged with pale brown.
DISTRIBUTION Central north-western Qld. **HABITAT AND HABITS** Occurs in tussock grassland dominated by Mitchell Grass *Astrebla* spp., on cracking clay soils. Shelters by day in cracks in soil, hunting nocturnally on the ground for insects and other arthropods.

Gilbert's Dunnart

■ *Sminthopsis gilberti* TL 155–184mm, including tail 75–92mm

DESCRIPTION Pale grey above, with white patch behind ear and white below, including feet, extending to below eye onto flanks. Tail thin and either slightly shorter (western) or same length (eastern) as head-body. Hindfeet slightly longer and tail proportionately

longer than those of Little Long-tailed Dunnart (see p. 39), and tail thicker and more pinkish in **White-tailed Dunnart** *S. granulipes*. **DISTRIBUTION** Inland south-western WA and small area around border in south-eastern WA and south-western SA. **HABITAT AND HABITS** Found in forests, woodland and heathland in west, and mallee shrubland in east. Nocturnal, probably sleeping by day in nest in dense vegetation or hollow log. Feeds on insects and other vertebrates.

Greater Hairy-footed Dunnart

■ *Sminthopsis hirtipes* TL 147–180mm, including tail 75–95mm

DESCRIPTION Sandy to brown above, with longer black guard hairs, and black patch around eye. Underparts white. Feet long and broad, and sparsely covered with silvery-white hairs above and below. Tail slightly longer than head-body, occasionally swollen at base and

with fine, pinkish-white fur. Larger than Lesser Hairy-footed Dunnart (see p. 44), with more hairy soles. **DISTRIBUTION** Arid to semi-arid mainland, from coastal western and south-eastern inland WA, to north-western SA and south-western NT. **HABITAT AND HABITS** Occurs in woodland, shrubland and spinifex grassland, generally with soft, sandy soil. Active at night, feeding on insects, spiders and small lizards, and shelters at other times in disused burrows of other animals, mainly spiders.

▪ DASYURIDS ▪

White-footed Dunnart
▪ *Sminthopsis leucopus* TL 140–200mm, including tail 70–90mm

DESCRIPTION Sandy above, with longer black guard hairs and patch of brown skin on snout. Whitish below, becoming more greyish on flanks, and feet pink, sparsely covered with white hairs. Tail similar in length or slightly shorter than head-body, and not swollen. Often confused with Common Dunnart (see p. 42). **DISTRIBUTION** Coastal and near coastal south-eastern NSW, southern Vic, Tas (including offshore islands), and uplands west of Cairns to Paluma, north-eastern Qld. **HABITAT AND HABITS** Occurs in variety of habitats in extensive range, including rainforests, cooler forests, woodland, scrubland, heathland and grassland. Sleeps by day in tree hollows, hollow logs or piles of bark, and hunts at night for terrestrial vertebrates, mainly small reptiles, and range of invertebrates.

Large Long-tailed Dunnart
▪ *Sminthopsis longicaudata* TL 260–310mm, including tail 180–210mm (Tjarrtjalaranpa; Yarntala; Yarrutju)

DESCRIPTION Generally grey above and off white to white below (including feet), with flattened head, pointed snout and very long tail, covered in small scales and with finely tufted brush tip. **DISTRIBUTION** Arid western and central Australia, from central WA to south-western NT and north-western SA. **HABITAT AND HABITS** Found in shrubby spinifex grassland, appearing to prefer rugged, rocky areas. Agile climber, and long tail is likely to assist with balance when jumping between boulders. Nocturnal, possibly emerging at dusk, and feeding on variety of arthropods, mainly ants and beetles.

◼ DASYURIDS ◼

Stripe-faced Dunnart
◼ *Sminthopsis macroura* TL 155–200mm, including tail 80–100mm

DESCRIPTION Greyish-brown washed with yellowish on face, and with blackish longitudinal line on top of head. Underparts, including feet, white. Tail swollen at base,

and similar in length to head-body. **DISTRIBUTION** Found throughout arid and semi-arid mainland, from western WA, through inland NT and SA, to central and western Qld and north-western NSW. **HABITAT AND HABITS** Occurs in woodland, shrubland and tussock grassland. Sleeps by day in soil cracks, and forages on the ground for insects and other arthropods. Tail used as fat storage.

Common Dunnart
◼ *Sminthopsis murina* TL 133–190mm, including tail 68–90mm

DESCRIPTION Light brown to mouse grey fur on the back, darker on the head and neck. Underparts are a light brown to almost white colour. Feet are pink flesh colour with white hairs. Quite large rounded ears, bulging eyes and a thin tail about the same size or slightly shorter than the body. Tail is swollen at the base. Has a pointed nose and long hind foot. Smaller and more delicate than the Antechinuses. **DISTRIBUTION** North-eastern Qld from Cooktown and inland Qld. Coast to inland NSW, and from Fleurieu Peninsula SA, inland through north-western Vic. **HABITAT AND HABITS** Prefers areas with grass cover within woodlands, dry open forests and heathlands, and ecotones of tropical moist forests.

Logs and rocks lined with leaves and grasses in crevices are key nesting requirements. Sleeps by day and feeds at night on a wide range of terrestrial invertebrates, including beetles, roaches, crickets and spiders. Can become torpid in winter and go into hibernation when the temperature drops below 15°C; a survival technique during unfavourable conditions.

◾ Dasyurids ◾

Ooldea Dunnart
◾ *Sminthopsis ooldea* TL 115–173mm, including tail 60–93mm

DESCRIPTION Pale to dark greyish-brown above, washed with yellow, and with blackish longitudinal line on forehead and patches around eyes. Underparts, including feet, white, extending onto lower cheeks and flanks; tail long and thin. **DISTRIBUTION** Arid areas of central Australia, including central and eastern WA, south-western NT and north-western SA. **HABITAT AND HABITS** Occurs in woodland, shrubland and tussock grassland in arid sandy or stony areas. Nocturnal, hunting for insects on the ground. Sleeps by day in shallow burrow.

Red-cheeked Dunnart
◾ *Sminthopsis virginiae* TL 170–270mm, including tail 87–135mm

DESCRIPTION Dark grey above, each hair having paler tip, and with reddish wash on forelegs and flanks. Head has reddish-brown cheeks and black longitudinal stripe on forehead. Underparts white to greyish-fawn. Tail thin and pinkish, and lightly furred with darker hairs (pale in some northern individuals). Australian females have 8 teats. **DISTRIBUTION** Kimberley region of WA, northern NT (including Melville Island) and eastern Qld, from northern Cape York Peninsula to around Rockhampton. Also New Guinea. **HABITAT AND HABITS** Found in margins of woodland, wetter grassland and swampland. Forages at night mainly for insects, but also small reptiles. .

DASYURIDS/ NUMBAT

Lesser Hairy-footed Dunnart
■ *Sminthopsis youngsoni* TL 128–141mm, including tail 62–70mm

DESCRIPTION Sandy yellowish-brown above, streaked with black. Darker on crown, and head has whitish stripe above eye and black along sides of snout to rear of eye. Tail pinkish, with slightly swollen base. White below, including broad feet, and soles of all feet covered in short hairs. Smaller than similar Greater Hairy-footed Dunnart (see p. 40), with proportionally shorter tail. **DISTRIBUTION** Arid inland of northern WA, southern NT and western Qld. **HABITAT AND HABITS** Occurs in sandy reddish deserts with grassy tussocks and spinifex. Nocturnal, foraging on the ground for insects and other invertebrates, and sleeping at other times in disused burrows.

MYRMECOBALIDAE, NUMBAT

Numbat ■ *Myrmecobius fasciatus* TL 360–490mm, including tail

DESCRIPTION Reddish-brown above, becoming darker brown to black with white transverse bars on rump, and head paler on sides with conspicuous blackish eye-stripe. Tail long, bushy and upcurved. **DISTRIBUTION** Occurs in south-western WA, and introduced to protected sites in south-eastern SA and western NSW. **HABITAT AND HABITS** Found in forests and woodland. Active by day, foraging mainly for termites, which are collected using long, sticky tongue, and sheltering at night in hollow logs, burrows or hollows in trees.

Adult   *Female carrying young*

BANDICOOTS

PERAMELIDAE, BANDICOOTS

Long-nosed Echymipera

■ *Echymipera rufescens* TL 375–500mm, including tail 80–100mm
(Rufous Spiny Bandicoot)

DESCRIPTION Reddish-brown above and on sides, becoming more blackish on top of head, snout and rump. Has a flecked appearance, and long, stiff black guard hairs protrude through undercoat (ground hair). Cream to buff below. Tail short, blackish-brown above and buff below, and sparsely furred. Four pairs of upper incisors, as opposed to the five pairs in other Australian bandicoots. Ears longer and more rounded than the Northern Long-nosed Bandicoot (see p. 48), which is also paler, particularly on tail. **DISTRIBUTION** Occurs on Cape York Peninsula, Qld, north from McIlwraith Range, and PNG and Indonesia. **HABITAT AND HABITS** Occurs in rainforests and grassy woodlands where it feeds at night on a range of fruits, seeds, nuts, leaves, fungi and invertebrates. Rests during the day in a shallow burrow of around 0.5m deep with two openings.

Northern Brown Bandicoot

■ *Isoodon macrourus* TL 380–695mm, including tail 80–215mm

DESCRIPTION Largest Australian bandicoot. Larger than similar Southern Brown Bandicoot (see p. 46) and **Golden Bandicoot** *I. auratus*. Fur is brindle-brown above, black and tan on dorsal side, heavily streaked with black and yellowish-buff; cream below and on feet. Snout long and pointed, ears shorter and more rounded than Northern Long-nosed Bandicoot (see p. 48); tail is short and bicoloured with sparsely furred dark grey-brown above. **DISTRIBUTION** Occurs in two geographically separated regions, the Kimberley WA, and through northern NT, northern and eastern QLD and north-eastern NSW. **HABITAT** Found in rainforests, wet forests, woodland, grassland and suburban gardens. Sleeps by day in a shallow scrape in the ground, covered with ground litter, or in hollow logs, emerging at night to forage through leaf litter and topsoil for insects (mainly) and other invertebrates, fruits and seeds.

◾ BANDICOOTS ◾

Southern Brown Bandicoot
◾ *Isoodon obesulus* TL 360–500mm, including tail 80–130mm
(Quenda)

DESCRIPTION Appears greyish-brown above from a distance, but is more greyish with prominent golden-brown streaks when seen at close range. Underparts cream to yellowish-grey. Tail short, brownish above and creamish-yellow below, and head has pointed snout and small, rounded ears. **DISTRIBUTION** South-western WA (including Daw Island), south-eastern SA (including Kangaroo Island and Nuyts Archipelago), southern Vic,

south-eastern NSW, ACT and Tas (including Bruny, Three Hummock and Sister Islands). **HABITAT AND HABITS** Found in forests, wetlands and moist heathland with dense vegetation. Sleeps day or night in nest placed under vegetation, or occasionally in disused burrows of other animals, emerging to forage in ground litter or dig in topsoil for variety of invertebrates, fungi, and plant roots and tubers.

Shark Bay Bandicoot ◾ *Perameles bougainville*
TL 275–420mm, including tail 60–100mm (mature females larger than males)
(Western Barred Bandicoot)

DESCRIPTION Pale grey to brownish above, with alternating paler and darker transverse bars on rump (less prominent in some parts of range), and whitish below. Tail short, and

head has large ears and long, pointed snout. **DISTRIBUTION** Species now confined to Bernier, Dorre and Faure Islands (introduced) in Shark Bay WA, and Roxby Downs SA. **HABITAT AND HABITS** Found in most vegetation types in limited range, with preference for vegetated coastal dunes. Shelters by day in nest of dried vegetation placed in saucer-shaped scrape in the ground, concealed under low-lying shrub, and forages alone for insects and plant material, supplemented with small vertebrates.

▪ Bandicoots ▪

Eastern Barred Bandicoot
▪ *Perameles gunnii* TL 340–460mm, including tail 70–110mm

DESCRIPTION Yellowish-brown above, streaked with silvery-white, extending onto base of white tail, and with 3–4 paler bars on rump. Greyish below. **DISTRIBUTION** Southern Vic, where has been reintroduced into predator-proofed reserves in Mount Rothwell and Hamilton Community Parklands, and into the wild on Phillip and French Islands. Also in Tas, including Bruny and Maria (introduced) Islands. **HABITAT AND HABITS** Occupies moist areas in grassy woodland and grassland, with combination of dense vegetation for shelter during the day and open areas for foraging at night. Feeds by digging in topsoil for wide variety of invertebrates, plant matter and fruiting bodies of fungi. Sleeps at other times in nest lined with grass.

Southern Long-nosed Bandicoot
▪ *Perameles nasuta* TL 450–700mm, including tail 125–155mm

DESCRIPTION Greyish-brown above, streaked with paler brown, and creamish-white below, with short, sparsely furred tail, long, pointed snout and moderately long, pointed ears. Upper surfaces of feet cream, with long, powerful claws for digging. Distinguished from the smaller Southern Brown Bandicoot (opposite) by its darker tail, and longer ears and snout. **DISTRIBUTION** Occurs along the coast and either side of Great Dividing Range near inland eastern Australia, from Mackay Qld, through eastern NSW to southern Vic. **HABITAT AND HABITS** This ground-dwelling mammal is found in rainforests, wet forests, dry open forests, heath, damp grassland and suburban gardens, where it feeds at dusk and through the night on invertebrates, occasionally very small vertebrates, fungi and underground tubers, which it finds by digging through the topsoil. Shelters by day in a shallow burrow, lined with leaves.

BANDICOOTS/BILBIES

Northern Long-nosed Bandicoot
■ *Perameles pallescens* TL 330–600mm, including tail 120–150mm

DESCRIPTION Greyish-brown above and whitish below, with a short, sparsely furred tail, long, pointed snout and moderately long, pointed ears. Northern Brown Bandicoot (see p. 45) has shorter, more rounded ears and prefers drier forests. **DISTRIBUTION** Coastal ranges of north-eastern Qld in two separated areas, including north-eastern Cape York Peninsula, and between Cooktown and Townsville. **HABITAT AND HABITS** Naturally occurring in rainforests and wet sclerophyll forests, where it is active at night feeding predominantly on insects, supplemented with plant matter and fungi, which are found by ripping apart rotting logs and branches. Sleeps at other times in a shallow burrow or under fallen timber.

THYLACOMYIDAE, BILBIES

Greater Bilby ■ *Macrotis lagotis* TL 500–840mm, including tail 200–290mm
(Dalgyte)

DESCRIPTION Fur soft, bluish-grey above, washed with rufous on sides and white below (including feet). Long, bicoloured tail, black on base and white towards tip, with extreme tip naked. Ears long and erect, and snout sharply pointed. **DISTRIBUTION** Natural range now reduced to central and north-western WA, through central and southern NT, to western Qld. Introduced and reintroduced to several mainland reserves in NSW, WA, SA and Qld, and Thistle Island, SA. **HABITAT AND HABITS** Found in hummock grassland and arid shrubland. Shelters by day (and at times during the night) in deep underground burrows, and feeds by digging in topsoil for variety of invertebrates and plant material.

KOALA/WOMBATS

PHASCOLARCTIDAE, KOALA

Koala ■ *Koala Phascolarctos cinereus*
TL 680–820mm

DESCRIPTION Has thick greyish fur (shorter and paler in north Australia, and longer and more grey brownish in south); round, woolly ears, and a prominent nose, which is smooth, black and vertically oval. Long, sharp claws adapted to climbing trees and no tail. **DISTRIBUTION** Found in eastern Australia, from south of Cape York Peninsula Qld, through central and eastern NSW, ACT and Vic (and along Murray River), to far south-eastern SA, and has been introduced to Phillip Island Vic and Kangaroo Island SA. **HABITAT AND HABITS** Inhabits eucalypt forests and woodland in lowland areas and along river systems, and can be seen along rainforest fringes. Arboreal. Walks on all fours along the ground between trees. Preferred food source of about ten different Eucalyptus species across its known habitat range in Australia, which also determines the home range for a group of Koalas as they feed and socially interact within a set territory. The young Koala is not born with the gut bacteria needed to digest eucalypt leaves, and obtains these by eating its mother's pap (faeces).

VOMBATIIDAE, WOMBATS

Northern Hairy-nosed Wombat
■ *Lasiorhinus krefftii*
TL 925–1,125mm, including tail 25mm

DESCRIPTION Stocky with short legs. Predominantly greyish-brown, streaked with darker brown or black. Acutely pointed, upright ears, and square snout with short hairs between large nostrils and long whiskers. **DISTRIBUTION** Restricted to Epping Forest region in central Qld. **HABITAT AND HABITS** Found in open eucalypt woodland. Feeds on the ground, mostly at night, on shoots of grasses and sedges, sleeping at other times in network of underground of burrows shared with several other individuals.

◼ Wombats ◼

Southern Hairy-nosed Wombat
◼ *Lasiorhinus latifrons* TL 795–965mm, including tail 25mm

DESCRIPTION Thickset body, with large head and short limbs. Fur soft and dense, whitish-grey to blackish above, and generally paler below. Upright, pointed ears and fleshy nosepad with short fur between large nostrils and long whiskers. **DISTRIBUTION** Arid to semi-arid areas of far south-western NSW, southern SA and near border in far south-western WA. **HABITAT AND HABITS** Occurs in open woodland and grassland. Active predominantly at night, feeding on the ground, mainly on shoots of native grasses. Sleeps at other times in network of underground burrows shared by up to 10 individuals in some populations, or in simple burrow that it occupies alone.

Bare-nosed Wombat
◼ *Vombatus ursinus* TL 725–1,175mm, including tail 25mm
(Common Wombat)

DESCRIPTION Stocky, with large, flattened head, short limbs and powerful claws for digging. Greyish-brown to blackish above and paler below, with bare nose and short, slightly rounded ears. **DISTRIBUTION** South-eastern mainland. From south-eastern Qld, through ranges of eastern NSW and ACT, to central Vic and Tas, including Flinders and Maria (introduced) Islands. Also in far south-eastern SA. **HABITAT AND HABITS** Found in forests, woodland and grassland. Sleeps by day in underground burrow and feeds on the ground, mostly at night, on native grasses, sedges and roots of shrubs.

BURRAMYIDAE, PYGMY-POSSUMS

Mountain Pygmy-possum
Burramys parvus TL 240–270mm, including tail 130–150mm (Burramys)

DESCRIPTION Greyish-brown above, generally with darker stripe on top of head and along back, and cream to grey below, with orange wash on underparts during breeding season. Tail thin, pinkish-grey and mostly naked, with limited covering of short hair at base. Ears rounded and snout moderately blunt. **DISTRIBUTION** Alpine and subalpine mountain ranges (above 1,200m) in south-eastern NSW and north-eastern Vic. **HABITAT AND HABITS** Found in rocky areas in alpine shrubland and heathland. Seasonally nocturnal, foraging on the ground for invertebrates, particularly moths, and vegetable matter such as seeds, fruits and nectar. Hibernates from onset of colder weather until snow melt, but may be active under snow drifts.

Long-tailed Pygmy-possum
Cercartetus caudatus TL 230–260mm, including tail 130–150mm

DESCRIPTION Greyish-brown above, paler below, with black eye-patches and long tail, thickened and tapering, with furry base. Snout pointed and ears rounded. **DISTRIBUTION** North-eastern Australia, from around Cooktown, south to around Townsville Qld.

HABITAT AND HABITS Occurs in rainforests and adjacent drier forests, and coastal plains. Mostly arboreal, feeding on insects, nectar and pollen. Sleeps by day in dome-shaped nest of leaves, placed in tree hollow or small crevice. Tail is prehensile, used for extra support while climbing and feeding.

PYGMY-POSSUMS

Western Pygmy-possum
■ *Cercartetus concinnus* TL 140–200mm, including tail 70–95mm

DESCRIPTION Grey to pale brown or reddish-brown above and white below, with prehensile tail furred at base and remainder naked, and having covering of fine scales. Head has blunted snout, long whiskers and large ears. Tail similar in length to head-body. **DISTRIBUTION** Southern Australia, from south-western WA, through southern SA

(including Kangaroo Island), to western Vic. **HABITAT AND HABITS** Occupies dry sclerophyll forests and mallee heathland. Sleeps by day in nest of leaves and shredded bark in tree hollow or smaller crevice, and forages after dusk mainly arboreally, for nectar and small invertebrates.

Little Pygmy-possum
■ *Cercartetus lepidus* TL 120–145mm, including tail 60–75mm

DESCRIPTION Pale grey, tinged with pale brown above, and pale greyish below, with hairs having grey tips with white bases, and white on throat. Large, rounded ears and sparsely furred, prehensile tail, which is often swollen at base. **DISTRIBUTION** South-eastern SA (including Kangaroo Island), north-western Vic and Tas. (DNA testing suggests

that Tas population may be a separate species.) **HABITAT AND HABITS** Occurs in semi-arid habitats on deep, sandy soils, including woodland, shrubland and heathland on mainland, and wetter forests in Tas. Feeds at night on range of small invertebrates and plant matter, especially nectar and pollen of *Banksia* flowers, and sleeps during the day.

PYGMY-POSSUMS/PETAURIDS

Eastern Pygmy-possum
- *Cercartetus nanus* TL 145–215mm, including tail 75–105mm

DESCRIPTION One of the smallest possums in the world. Pale brown to grey above and pale greyish below, with white tips on hairs. Tail brownish, almost naked and prehensile. The Little Pygmy-possum (opposite), with which it overlaps in range in the south, is similar, but smaller and less than half the weight. **DISTRIBUTION** Occurs within south-eastern Australia, from far south-eastern Qld, through eastern NSW, ACT and southern Vic, to south-eastern SA and Tas (including islands of Bass Strait). Also occurs west of the Great Dividing Range on the western slopes NSW to the Pilliga and down to Wagga Wagga. **HABITAT AND HABITS** Inhabits rainforests, forests, woodland and heathland. Feeds largely on nectar and pollen, supplemented with insects and spiders. Shelters by day in a nest of shredded bark and leaves placed in a hollow or under bark. Emerges shortly after dusk to forage both in trees and on the ground. The species goes into torpor in winter, with body curled, ears folded and a body temperature close to surrounding environment.

PETAURIDAE, STRIPED AND LEADBEATER'S POSSUMS, AND LESSER GLIDERS

Torresian Striped Possum
- *Dactylopsila trivirgata*
TL 570–620mm, including tail 320–350mm

DESCRIPTION White head and body with conspicuous contrasting black to greyish-brown longitudinal stripes along sides and top of head, top of body and down legs. Ears black, and tail long, bushy and mostly blackish with white tip. Fourth digit of front feet longer than others. **DISTRIBUTION** Coastal and adjacent ranges of north-eastern Qld. **HABITAT AND HABITS** Found in rainforests, wet sclerophyll forests and adjacent woodland. Nocturnal, foraging mostly in trees, mainly for insects, supplemented with nectar and fruits. Extracts wood-boring larvae from branches by chewing away surface and using its enlarged finger as a probe. Sleeps at other times in tree hollow lined with leaves, or in tree ferns.

PETAURIDS

Leadbeater's Possum
- *Gymnobelideus leadbeateri*
TL 295–350mm, including tail 145–180mm
(Fairy Possum)

DESCRIPTION Grey to brownish above, with darker blackish vertebral stripe, and cream-white below. Tail long and thickened towards tip. Lacks gliding membrane, which is present in similar looking Sugar Glider (see opposite). DISTRIBUTION Confined to a few fragmented areas, mainly above 500m, in central Vic. HABITAT AND HABITS Inhabits tall mixed eucalypt and wattle forests with dense shrubby understorey, lowland swamps and subalpine woodland. Sleeps by day in small groups in tree hollows lined with shredded bark, emerging at dusk to feed arboreally on tree exudates, invertebrates and nectar. Key population threats include collapse of large hollow-bearing trees, bushfires and logging.

Savanna Glider
- *Petaurus ariel* TL 300–450mm, including tail 160–235mm

DESCRIPTION Brownish-grey above, with a blackish stripe running from between eyes and along the back, typically to a point between the hind legs. Warm cream below. Tail blackish and bushy, snout pointed, ears broad and eyes black, large, prominent and

protruding. Large gliding membrane attached at wrists and ankles. Similar in appearance to the Squirrel Glider (see p. 56), but smaller and nose more pointed. DISTRIBUTION Widely distributed through northern Australia, from the Kimberley region, WA, through northern NT, to northwestern Qld. Also on several offshore islands. Species split from Sugar Glider (opposite) and Krefft's Glider (see p. 56). HABITAT AND HABITS Inhabits wooded savannas, where it feeds nocturnally on a range of foods including sap, nectar, pollen, insects, arachnids and small vertebrates. Individuals use tree hollows for refuge during the day and breeding. Population sizes of this and other similar-sized mammals in the region are negatively impacted by intense bushfires.

PETAURIDS

Yellow-bellied Glider
Petaurus australis TL 690–800mm, including tail 430–480mm

DESCRIPTION Grey above, with conspicuous black vertebral stripe, and white (young) or yellow (adults) below. Feet black, tail long and black, edged with grey at base. Gliding membrane large and edged with black.
DISTRIBUTION Occurs in eastern Australia, from northeastern Qld, through eastern NSW, ACT, eastern and southern Vic, to far southeastern SA. **HABITAT AND HABITS** Inhabits mainly mature wet eucalypt forests in an ecotone between rainforest and drier woodland ecosystems in the temperate, sub tropics and tropical areas. In the south-east, it can occur in dry open eucalypt forests. Sleeps by day in pairs or small family groups in tree hollows, and feeds in trees at night on sap, nectar, pollen, honeydew and invertebrates, gliding between trees as it forages within its large home range. Feeds on sap from various trees such as Flooded Gum *Eucalyptus grandis*, which is a major food source, by tapping into the bark and making 'V' notches.

Sugar Glider
Petaurus breviceps TL 295–395mm, including tail 140–220mm

DESCRIPTION Brownish-grey to blue-grey above, with a blackish stripe running from between eyes to centre of back, and cream to greyish below. Tail is blackish and bushy, occasionally tipped with white, snout is short and rounded, ears are broad. Large gliding membrane (edged with blackish and white) is attached at wrists and ankles (able to glide up to 90m). Black beady eyes are large, prominent and protruding. **DISTRIBUTION** Southeastern Qld and far eastern NSW. It was recently split from a wider ranging species complex, which also included the Savanna Glider (opposite) and the Krefft's Glider (see p. 56). **HABITAT AND HABITS** Inhabits rainforests, wet and dry sclerophyll forests and woodland. Nocturnal, sheltering by day in a hollow tree, lined with leaves, and feeds arboreally on nectar, pollen, honeydew and arthropods. Individuals chew the bark from trees to feed on sap when other food sources are rare.

▪ Petaurids ▪

Squirrel Glider
▪ *Petaurus norfolcensis* TL 390–540mm, including tail 220–300mm

DESCRIPTION Brownish-grey to blue-grey above, with distinct blackish stripe running from centre of head to middle of back, and white to cream below, becoming dark yellowish in older individuals in northern parts of range. Tail long and wide, with bushy black fur, snout moderately pointed, ears narrow and gliding membrane attached at wrists and ankles. **DISTRIBUTION** Coast, ranges and slopes of eastern Australia from northern Qld, through eastern NSW, to inland central Vic. **HABITAT AND HABITS** Occurs in rainforests, open forests and woodland. Forms small social groups, sheltering by day in tree hollows, typically with small entrance holes. Emerges at night to feed arboreally on nectar, pollen, honeydew, sap and small invertebrates. Feeding on small vertebrates and birds' eggs has been recorded.

Krefft's Glider
▪ *Petaurus notatus* TL 330–420mm, including tail 175–230mm

DESCRIPTION Brownish-grey to blue-grey above, with a well-defined blackish dorsal stripe running from between eyes to centre of back, and cream to greyish below. Tail bushy, strongly attenuated and blackish with white dorsal stripe and tip. Snout short and rounded, ears broad and large. Gliding membrane (edged with blackish and white) attached at wrists and ankles. Eyes, black, large and bulging. **DISTRIBUTION** Widespread through eastern Australia, from northern Qld, through NSW, Vic and Tas, to far south-eastern SA. Species split from Sugar Glider (see p. 55) and ranges may overlap. **HABITAT AND HABITS** Occurs in a wide range of habitats through its large range, from rainforests to dry woodlands. Omnivorous and feeds nocturnally on a range of foods including sap, nectar, pollen, insects, arachnids and small vertebrates. Uses tree hollows for refuge during the day and breeding.

PSEUDOCHEIRIDAE, RING-TAILED POSSUMS AND GREATER GLIDERS

Central Greater Glider

Petauroides armillatus TL 700–950mm, including tail 400–530mm

DESCRIPTION Generally greyish, darker on crown and along middle of back, paler on sides and with large, rounded ears. Large gliding membrane between ankles and elbows, and long, bushy tail. **DISTRIBUTION** Eastern Qld, inland to around Roma, and from around Yeppoon in south to Eungella in north. **HABITAT AND HABITS** Found mainly in eucalypt forests and woodland with abundant large tree hollows for shelter. Feeds at night almost exclusively on young leaves and buds, supplemented with flowers. Arboreal, moving between trees in small home range by gliding or through connected parts in the canopy.

Northern Greater Glider

Petauroides minor TL 700–880mm, including tail 400–480mm

DESCRIPTION Generally greyish, with darker longitudinal dorsal stripe on crown and tail. Paler on sides, and with moderately large, rounded ears. Large gliding membrane between ankles and elbows. Tail long and bushy. **DISTRIBUTION** North-eastern Qld, from Mount Windsor Tableland to around Dawson River, and isolated populations in Gregory Range and Einasleigh Uplands. **HABITAT AND HABITS** Found mainly in eucalypt forests and woodland with abundant large tree hollows for shelter. Feeds at night almost exclusively on young leaves and buds, supplemented with flowers. Arboreal, moving between trees in small home range by gliding or through connected parts in the canopy.

◾ Pseudocheirids ◾

Southern Greater Glider
◾ *Petauroides volans*
TL 800–1050mm, including tail

DESCRIPTION Variable colouration above, from uniformly cream to dark grey, or combination of both. Whitish below, with large, furry ears, short snout and pale spot behind ears. Tail long and bushy (not prehensile), and gliding membrane attached at elbows, knees and ankles. **DISTRIBUTION** Central and eastern Vic, eastern NSW and south-eastern Qld. **HABITAT AND HABITS** Found mainly in eucalypt forests and woodland with abundant large tree hollows for shelter. Feeds at night almost exclusively on young leaves and buds, supplemented with flowers. Arboreal, moving between trees in small home range by gliding or through connected parts in the canopy.

Western Ring-tailed Possum
◾ *Pseudocheirus occidentalis* TL 600–800mm, including tail 300–400mm (Ngwayir; Womp)

DESCRIPTION Uniformly dark brownish-grey above and creamish-white below, with pointed snout and rounded, mostly naked ears. Tail long and prehensile, with short fur, and either mostly creamish-white or just with small creamish-white tip. **DISTRIBUTION**

Fragmented areas in far south-western WA, between Bunbury area to near Albany. Reintroduced, mostly unsuccessfully, into parts of former range. **HABITAT AND HABITS** Occupies woodland. Arboreal and largely solitary, sleeping by day in tree hollows, in dreys or dense vegetation, or in hollow logs or stumps, and occasionally entering roof spaces in urban areas. Feeds at night on leaves, flowers and fruits.

◾ Pseudocheirids ◾

Eastern Ring-tailed Possum
◾ *Pseudocheirus peregrinus* TL 600–760mm, including tail 300–380mm

DESCRIPTION Pale grey on back with white patches behind eyes, ears and under neck, and orange to brown tinges on limbs and tail. Belly white. Long prehensile tail, one third white on its tip. **DISTRIBUTION** Widely distributed in eastern Australia, from Cape York Peninsula Qld, through eastern NSW, ACT and Vic, to south-eastern SA (including Kangaroo Island), and Tas (including islands of Bass Strait). **HABITAT AND HABITS** Inhabits rainforest, sclerophyll forests, woodland, dense understorey and scrub. Predominantly arboreal, using its prehensile tail like a fifth limb to climb from branch to branch in the forest. Forefoot structure, with a gap between second and third fingers, also allows it to hold onto branches securely. Feeds at night on leaves, flowers and fruits, and shelters during the day in nests constructed of sticks or grass in tree holes, tree forks or dense vegetation.

Daintree River Ring-tailed Possum
◾ *Pseudochirulus cinereus* TL 650–760mm, including tail 320–395mm

DESCRIPTION Pale brown to dark brown above, with darker brown or blackish longitudinal stripe from centre front of head to lower back, and creamish-white below. Tail long and prehensile, with long, tapering white tip and exposed pinkish skin on underside. **DISTRIBUTION** Small, discontinuous range, within Mount Carbine and Mount Windsor Tablelands and Thornton Peak, north-eastern Qld. **HABITAT AND HABITS** Found in rainforests above 420m. Arboreal and mostly solitary, sleeping by day in tree hollows or among epiphytes, and emerging at night to feed almost exclusively on leaves, supplemented with some fruits.

PSEUDOCHEIRIDS

Herbert River Ring-tailed Possum
■ *Pseudochirulus herbertensis* TL 625–770mm, including tail

DESCRIPTION Adults uniformly blackish, normally with white markings on underside, top of forearm and tip of long, tapering, prehensile tail. Juveniles pale brown, with darker longitudinal stripe from top of head to upper back. Snout pointed, ears small and eyes with narrow, fleshy eye-ring.

DISTRIBUTION North-eastern Qld, from Atherton Tablelands, south to Herbert River. **HABITAT AND HABITS** Occurs in rainforests above 200m. Arboreal and mostly solitary, sleeping by day in tree hollows, among epiphytes or within dreys, and emerging at night to feed almost exclusively on leaves, supplemented with flowers and fruits.

Rock Ring-tailed Possum
■ *Petropseudes dahlii* TL 530–755mm, including tail 200–270mm
(Wogoit)

DESCRIPTION Grey above, heavily flecked with reddish and silverish-white, with blackish longitudinal stripe from between eyes to middle of back. Whitish-grey below, with reddish-orange staining around sternal and tail glands. Body compact, and tail furred at base and naked for terminal half.

DISTRIBUTION Inland of northern WA, central northern NT (and Groote Eylandt) and north-western Qld. **HABITAT AND HABITS** Found in rocky outcrops in woodland. Feeds at night in pairs or small family groups, among rocky areas or in trees, on plant material, including leaves, flowers and fruits. Sleeps at other times in rocky crevices.

PSEUDOCHEIRIDS/HONEY POSSUM

Green Ring-tailed Possum
■ *Pseudochirops archeri*
TL 610–750mm, including tail 310–370mm

DESCRIPTION Greenish-grey above, flecked with yellow, silver and black, with two faint grey longitudinal stripes along centre back, each bordered with brownish-grey, and white below. Face grey, with conspicuous white patches under eyes and behind ears. Tail moderately short and thickened at base, with longer fur along upper surface, and white terminal third. DISTRIBUTION North-eastern Qld, including Mount Carbine, Mount Windsor and Atherton Tablelands, Paluma and Seaview Ranges and Mount Molloy. HABITAT AND HABITS Occupies upland rainforests down to about 300m. Arboreal and mostly solitary, sleeping by day in foliage and on branches, and feeding at night almost exclusively on leaves, supplemented with some fruits.

TARSIPEDIDAE, HONEY POSSUM

Honey Possum ■ *Tarsipes rostratus* TL 135–185mm, including tail 70–100mm
(Noolbenger)

DESCRIPTION Pale brown to greyish above, with dark vertebral stripe from nape to base of tail, with 2 parallel paler brown stripes. Underparts creamish-white. Snout greatly elongated, with numerous black whiskers. Ears rounded, eyes towards top of head and tail thin and mostly naked. DISTRIBUTION Coastal and near coastal south-western WA, from Kalbarri to around Madura. HABITAT AND HABITS Occurs in complex sandplain heathland with a variety of flowering plants. Mostly nocturnal, foraging for nectar and pollen both arboreally and terrestrially, and sheltering at other times in tree hollows, crevices or disused birds' nests. Relies on continuous access to flowering plants, but will enter torpor in colder months, when food abundance is lower.

ACROBATIDAE, FEATHER-TAILED GLIDERS
Broad-toed Feather-tailed Glider ■ *Acrobates frontalis*
Narrow-toed Feather-tailed Glider
■ *A. pygmaeus* TL 135–180mm, including tail 70–100mm

DESCRIPTION Smallest gliding mammals. Greyish above and creamish-white below. Tail prehensile and 'feather-like', with long, stiff black hairs along either side (*frontalis* has naked tip). Tips of third and fourth toes of hindfoot broader and heart shaped in *frontalis*, and narrow and rectangular in *pygmaeus*. **DISTRIBUTION** Coast and ranges of eastern and south-eastern Australia (including Fraser Island). Boundaries in distributions of species unclear, although *frontalis* is the only species in north and Fraser Island. **HABITAT AND HABITS** Found in forests and woodland. Omnivorous, feeding nocturnally in trees in small family groups on nectar, sap and insects.

Broad-toed Feather-tailed Glider   *Narrow-toed Feather-tailed Glider*

PHALANGERIDAE, CUSCUSES AND BRUSH-TAILED POSSUMS
Southern Common Cuscus
■ *Phalanger mimicus* TL 350–400mm, including tail 280–350mm

DESCRIPTION Grey-brown above, with darker brown stripe from top of snout to rump (more distinct on head and neck), and whitish below. Males have yellowish wash to chest and sides of neck. Snout slightly pointed, and ears naked and rounded. Fur dense and woolly, and tail prehensile, with terminal two-thirds naked and dark pinkish. **DISTRIBUTION** Eastern Cape York Peninsula, northern Qld, in vicinity of Iron and McIlwraith Ranges. **HABITAT AND HABITS** Inhabits lowland rainforests. Mostly arboreal and generally solitary, sleeping by day in dens or tree hollows, and feeding at night on leaves, flowers and fruits. Less diurnal than Australian Spotted Cuscus (see opposite).

PHALANGERIDS

Australian Spotted Cuscus
■ *Spilocuscus nudicaudatus*
TL 660–875mm, including tail 310–435mm

DESCRIPTION Generally greyish, with white neck and underparts, and males normally blotched grey and white, but occasionally white with yellow wash on rump. Face rounded, snout short, ears small and mostly hidden, and eyes facing forwards. Fur dense and woolly, and tail prehensile, with terminal two-thirds naked and yellowish-pink to reddish. **DISTRIBUTION** Cape York Peninsula, northern Qld. **HABITAT AND HABITS** Occupies rainforests, open forests, woodland and mangroves. Arboreal and mostly solitary, sleeping by day on branches in the canopy, and feeding at night on leaves, flowers, fruits and some invertebrates.

Short-eared Brush-tailed Possum ■ *Trichosurus caninus*
Mountain Brush-tailed Possum
■ *T. cunninghami* TL 600–920mm, including tail 250–400mm (Bobuck)

DESCRIPTION Usually dark silvery-grey above, occasionally tinged with reddish-brown, and pale cream or white below, but some individuals entirely blackish. Tail black and bushy, tapering towards tip, with section of naked skin towards tip on underside. Ears short and rounded (shorter in *caninus*). **DISTRIBUTION** Eastern Australia. *T. caninus* from around Gladstone Qld, south to around Ulladulla NSW, and *T. cunninghami* from around Ulladulla NSW to central Vic. **HABITAT AND HABITS** Occurs in variety of wooded habitats, including tall, open and closed forests, and adjacent pine plantations. Sleeps by day in tree hollows or rocky crevices.

T. cunninghami

T. caninus

▪ Phalangerids ▪

Common Brush-tailed Possum
▪ *Trichosurus vulpecula* TL 600–950mm, including tail 250–400mm
(Koomal)

DESCRIPTION Usually silvery-grey above and pale grey or white below, but occasionally entirely blackish or rich orange. Males have dark orange wash on chest from scent marking. Tail prehensile and normally bushy, but can be lightly haired, tapering towards tip, with section of naked skin on underside. **DISTRIBUTION** Western and south-western WA, northern NT, eastern Qld, eastern and southern NSW, ACT, Vic, Tas and south-eastern SA. **HABITAT AND HABITS** Found in wooded habitats throughout range. Shelters in large tree hollows and roof spaces in suburban dwellings, and occasionally in underground burrows. Nocturnal and mostly arboreal, feeding on range of plant material, including leaves, fruits and flowers, human food scraps, invertebrates and vertebrates.

Scaly-tailed Possum
▪ *Wyulda squamicaudata* TL 610–700, including tail 300mm
(Wyulda; Yilangal; Ilangurra)

DESCRIPTION Grey flecked with black, with blackish stripe from shoulders to rump, and off white below. Tail prehensile, with reddish-orange fur at base and scaled on the remainder. **DISTRIBUTION** Kimberley region of north-western WA. **HABITAT AND HABITS** Found in forests, woodland and vine thickets, with areas of exposed sandstone boulders. Shelters by day usually in crevices in or underneath rocks, emerging at night to feed mainly in trees on leaves, flowers, fruits and seeds, presumably supplemented with some invertebrates.

HYPSIPRYMNODONTIDAE, MUSKY RAT-KANGAROO

Musky Rat-kangaroo

■ *Hypsiprymnodon moschatus* TL 285–430mm, including tail 123–159mm

DESCRIPTION Reddish-brown flecked with darker brown. More greyish on head, and tail darker with scales. Unlike other kangaroos, has retained first toe on hindfeet and thus has five toes on each hindfoot.

DISTRIBUTION Coast and ranges of north-eastern Qld.

HABITAT AND HABITS Found in tropical rainforests. Shelters during the night and heat of the day in ball-shaped nest made from fallen leaves. Feeds at other times on vegetable matter, including fruits and seeds, and small animals. Gallops rather than hops.

POTOROIDAE, BETTONGS AND POTOROOS

Rufous Bettong

■ *Aepyprymnus rufescens* TL 710–780mm, including tail 338–390mm

DESCRIPTION Fur long and wiry, reddish-brown above, flecked with silverish-grey, and pale greyish below. Head has triangular, pointed ears, short, hairy snout and sparsely furred nosepad. Naked skin around eyes and inside ears pink to pinkish-orange.

DISTRIBUTION Formerly wider ranging (including along sections of Murray River), but now confined to eastern Qld and north-eastern NSW.

HABITAT AND HABITS Found in forests and woodland. Favours open areas or those with grassy understorey, with adjacent thicker ground vegetation. Sleeps by day in dome-shaped nest of woven fibrous material, placed in shallow scrape on the ground, and emerges at dusk to feed on grasses, herbs, roots and underground tubers.

◾ Bettongs & Potoroos ◾

Eastern Bettong ◾ *Bettongia gaimardi* TL 600–680mm, including tail

DESCRIPTION Yellowish-grey to brownish-grey above, heavily flecked with white. Terminal third of tail darker reddish-brown, with contrasting white tip (not visible in some individuals). Underparts whitish, often washed with grey. Nose naked, with naked area not extending onto snout. **DISTRIBUTION** Formerly on south-eastern mainland (up to early 1900s), but now confined to Tas. **HABITAT AND HABITS** Occurs in open forests and woodland with heath or grassy understorey. Shelters by day in small nest constructed under dense vegetation or fallen timber. Feeds at night on subterranean fungi, supplemented with roots, shoots, bulbs and seeds of plants, and possibly small invertebrates. Uses prehensile tail to carry grasses and bark to nest site.

Burrowing Bettong
◾ *Bettongia lesueur* TL 495–700mm, including tail 280–305mm
(Boodie)

DESCRIPTION Small and stocky. Grey above and paler grey below, with fat, sparsely haired tail, short snout and short, rounded ears. **DISTRIBUTION** Restricted to islands off north-western WA, including Barrow, Boodie and Faure Islands, and mainland sanctuaries at Heirisson Prong WA, Roxby Downs SA, Yookamurra SA and Scotia Sanctuary, NSW, where it has been reintroduced. **HABITAT AND HABITS** Found in rocky or sandy areas. Shelters in large communal burrows with many entrances, which it digs in the ground. Emerges at dusk to feed nocturnally on subterranean fungi, roots, bulbs, fruits and seeds of plants, and small insects, including termites. Island populations may also feed on turtles' eggs or hatchlings. Shelters in underground burrows.

Bettongs & Potoroos

Brush-tailed Bettong
■ *Bettongia penicillata* TL 590–740mm, including tail 290–360mm
(Woylie)

DESCRIPTION Grey to greyish-brown above and paler grey below. Distal half of tail has crest of blackish hairs. **DISTRIBUTION** Natural range now restricted to south-western WA. Also in conservation reserves in south-western WA, southern SA and western NSW.

HABITAT AND HABITS Occupies open forests and woodland with a dense understorey, and neighbouring areas of heathland. Shelters by day, occasionally with other individuals, in nest built on the ground. Emerges at night to feed solitarily on fruiting bodies of subterranean fungi, bulbs, roots, seeds and resin of plants, and presumably some insects.

Northern Bettong
■ *Bettonga tropica* TL 584–768mm, including tail 317–365mm

DESCRIPTION Dark grey above, including first two-thirds of black-tipped tail, and paler grey below. Nosepad naked, but remainder of snout furred. **DISTRIBUTION** North-eastern Qld; now restricted to Lamb Range and Mount Carbine Tableland, and possibly Coane Range and Mount Windsor Tableland (but no documented sightings in these latter locations since 2003). **HABITAT AND HABITS** Occupies margins of rainforests, tall sclerophyll forests and grassy woodland (800–1,200m). Mostly solitary, sheltering by day and emerging at night to forage for subterranean fungi, as well as roots, tubers and seeds of plants, and small invertebrates.

BETTONGS & POTOROOS

Gilbert's Potoroo
■ *Potorous gilbertii* TL 485–520mm, including tail 215–230mm
(Ngilgyte)

DESCRIPTION Greyish-brown above, with slender, downcurved snout and heavyset face. Naked patch of skin extends from nose to end of snout. Paler on underparts.

DISTRIBUTION Confined to Two Peoples Bay Nature Reserve, Waychinnup National Park (reintroduced) and Bald Island (introduced) in south-western WA. **HABITAT AND HABITS** Inhabits dense mature heathland. Found in small groups, and shelters during the day. Emerges at night to feed almost exclusively on variety of fungi.

Long-footed Potoroo
■ *Potorous longipes* TL 695–740mm, including tail 315–320mm

DESCRIPTION Greyish-brown above and grey below. Hindfoot longer than head, and with small, raised pad between heel and base of fused second and third toes.
DISTRIBUTION Fragmented range in far south-eastern NSW and eastern Vic. **HABITAT AND HABITS** Found in wet and dry forests with a dense understorey and loose (friable) soils. Mainly nocturnal, feeding predominantly on subterranean fungi, supplemented with plant material and small invertebrates.

Long-nosed Potoroo
■ *Potorous tridactylus* TL 580–650mm, including tail 160–230mm

DESCRIPTION Grey or brown above and paler below, with a naked patch of skin extending along long, pointed nose and onto snout. Hindfoot shorter than length of nose. Smaller, with proportionally smaller feet than Long-footed Potoroo (opposite). **DISTRIBUTION** Eastern and south-eastern Australia, from south-eastern Qld, through eastern NSW and southern Vic, to far south-eastern SA and Tas (including islands of Bass Strait). **HABITAT AND HABITS** Found in moist areas with dense ground cover within a range of habitats, including rainforests, wet sclerophyll forests, woodland, shrubland and heathland. Predominantly nocturnal and rests during day in groundcover of thick vegetation. In winter occasionally forages during day. Omnivorous and feeds mostly on subterranean fungi and roots, supplemented with plant material, insects, larvae, snails and other soft-bodied animals that live in the soil. Australia's original Truffle hunter.

MACROPODIDAE, KANGAROOS AND WALLABIES

Bennett's Tree-kangaroo
■ *Dendrolagus bennettianus* TL 1,130–1,590mm, including tail 630–940mm (Jarabeena)

DESCRIPTION Greyish-brown to brown above. More reddish-brown on shoulders, neck and crown, and becoming greyish on remainder of head, and with yellowish wash on snout. Upperside of tail blackish on base and black on brushy tip. **DISTRIBUTION** Coast and ranges (up to about 1,400m) of central eastern Cape York Peninsula Qld, from Daintree River to Annan River and Mount Windsor Tablelands. **HABITAT AND HABITS** Inhabits rainforests. Feeds mainly in the canopy on leaves, stems and shoots of trees, creepers and ferns, and some fruits, but also descends to forage on the ground. Uses long tail for balance when climbing in tree branches, but tail is not able to grip. Can leap to the ground from height of about 18m.

Kangaroos & Wallabies

Lumholtz's Tree-kangaroo
■ *Dendrolagus lumholtzi* TL 1,120–1,330mm, including tail 655–740mm (Boongary)

DESCRIPTION Blackish-grey above, with yellowish on neck, rear of head, crown and forehead, and blackish face. Tail blackish and brush tipped, flecked with reddish-brown on base. Underparts yellowish-brown, and feet black. **DISTRIBUTION** Coast and ranges of south-eastern Cape York Peninsula Qld, from Daintree River to Herbert River. **HABITAT AND HABITS** Inhabits rainforests and riverine vegetation in more open forests. Sleeps for most of the day curled up high in a tree, and feeds mainly at night on leaves, bark and some fruits. Although able to hop, generally walks on all fours along branches, using long tail to balance itself. Occurs in small, widespread groups, normally consisting of a single male and a number of females.

Allied Rock-wallaby
■ *Petrogale assimilis* TL 879–1,135mm, including tail 409–550mm

DESCRIPTION Colour variable, but generally greyish-brown to dark brown above, with paler streak on cheek, and darker patch near armpit. Tail has blackish, brushy tip; white tip on some individuals. Underparts and limbs yellowish-brown. **DISTRIBUTION** Eastern

Qld, from Townsville to Cairns and inland to Hughenden (south-west) and Croydon (north-west). Also Magnetic and Palm Islands. **HABITAT AND HABITS** Lives in colonies in rocky outcrops in range of open habitats. Mainly nocturnal, but may also be active by day, and shelters at other times in caves and rocky crevices. Feeds on variety of grasses, herbs, seeds, fruits and flowers.

▪ Kangaroos & Wallabies ▪

Western Short-eared Rock-wallaby
▪ *Petrogale brachyotis*
TL 831–1,070mm, including tail 370–565mm

DESCRIPTION Pale grey above, washed with reddish or yellowish on shoulders and hips, with black dorsal stripe from crown to upper back, and conspicuous silverish streaking on shoulders. Underparts greyish and feet black. Tail brownish, with dark, sometimes bushy tip. DISTRIBUTION Northern Australia, from Kimberley region WA, to western NT. HABITAT AND HABITS Lives in open, grassy woodland, favouring rocky outcrops, cliffs and gorges. Mostly nocturnal, but may also feed during cooler days on grasses, herbs, seeds, fruits and other plant material. Shelters at other times in caves and similar places.

Monjon Rock-wallaby
▪ *Petrogale burbidgei* TL 570–650mm, including tail 265–290mm
(Warabi)

DESCRIPTION Smallest rock-wallaby. Generally greenish-brown above and paler on flanks, with reddish-brown and blackish mottling. Face more reddish-buff, with large eyes and paler eye-stripe from snout to ear. Tail, brushy tipped, curled over back as it hops. Underparts creamish-yellow and paws greenish-grey. Narbalek Rock-wallaby (see p. 72) generally greyer with longer ears. DISTRIBUTION North-western Kimberley, and Bigge, Boongaree and Katers Islands WA. HABITAT AND HABITS Occupies rocky sandstone areas with high annual rainfall, and numerous rocky crevices and caves, mainly in open woodland. Feeds mostly at night on grasses and ferns.

Kangaroos & Wallabies

Nabarlek Rock-wallaby

■ *Petrogale concinna* TL 510–660mm, including tail 220–310mm (Nabarlek)

DESCRIPTION Dull reddish-brown above, washed with grey, occasionally with black stripe on shoulder, and tail gradually darkening towards brushed tip. Underparts greyish-white. (See also Monjon Rock-wallaby p. 71). **DISTRIBUTION** Tropical north-western Australia, from Kimberley region and Augustus, Borda, Hidden and Long Islands WA, to eastern Arnhem Land NT. **HABITAT AND HABITS** Lives in rocky areas. Shelters by day in caves, emerging at night to feed on variety of plant material, including grasses, ferns and herbs. Can also be seen feeding by day during cooler parts of the year. **Notes** Able to continually produce supernumerary molar teeth throughout its life.

Godman's Rock-wallaby

■ *Petrogale godmani* TL 975–1,210mm, including tail 480–640mm

DESCRIPTION Generally greyish above, with reddish-brown wash on back and base of tail, and blackish on shoulders and tail. Face and cheeks more yellowish-brown, with darker facial stripe. Underparts greyish-white, washed with yellowish-buff on chest. **DISTRIBUTION** Fragmented populations in southern Cape York Peninsula Qld. **HABITAT AND HABITS** Found in rocky areas in seasonally wet open forests. Shelters by day in caves and rocky crevices, emerging late in the day to forage. Hybridizes with Mareeba Rock-wallaby (see p. 74) where ranges overlap.

KANGAROOS & WALLABIES

Herbert's Rock-wallaby
Petrogale herberti TL 980–1,275mm, including tail 510–660mm

DESCRIPTION Greyish-brown above, more greyish on shoulder and reddish on rump, with black patch near shoulders, white stripe along sides and black dorsal stripe from front of head to upper back. Head greyish with obscure paler stripe on cheek, and ears have black bases. Underparts whitish to pale brown. **DISTRIBUTION** South-eastern Qld, from about 100km north-west of Brisbane, to around Rockhampton in north and Clermont in west. **HABITAT AND HABITS** Lives in colonies in boulder fields and rocky outcrops. Active day and night, feeding on variety of plant material, and sheltering at other times in caves and rocky crevices.

Unadorned Rock-wallaby
Petrogale inornata
TL 884–1,210mm, including tail 430–640mm
(Unadorned Rock Wallaby; Plain Rock Wallaby)

DESCRIPTION Colour variable, and relative to colouration of rocks in areas it inhabits. Generally greyish-brown above with paler cheek-stripe, and occasionally with short blackish dorsal stripe from forehead to upper back. Tail pale brown at base and blackish towards moderately brushed tip. Underparts and limbs yellowish-brown. **DISTRIBUTION** Eastern Qld, between Townsville and Rockhampton. Also Whitsunday Island. **HABITAT AND HABITS** Small colonies live in rocky outcrops and boulder piles. Mainly active at night, foraging on variety of plant material, and sheltering by day in rocky crevices and caves.

KANGAROOS & WALLABIES

Black-footed Rock-wallaby
■ *Petrogale lateralis* TL 853–1,160mm, including tail 407–605mm (Warru)

DESCRIPTION Generally dark brown streaked with silver above, more greyish on shoulders, and with white and dark brown stripe along sides. Head brownish on forehead and snout, with pale buff cheek-stripe and greyish on lower jaw and cheeks. Ears have dark brown tips and white bases. DISTRIBUTION Highly fragmented in arid southern NT, northern SA and eastern WA, but also isolated subpopulations in western and northern WA, including Barrow Island. Introduced to south-western WA and islands off southern SA. HABITAT AND HABITS Occupies rocky areas. Shelters by day in caves or deep crevices, emerging at dusk to feed on variety of plant material, including grasses, leaves, herbs, fruits and seeds.

Mareeba Rock-wallaby
■ *Petrogale mareeba* TL 840–1,078mm, including tail 415–530mm

DESCRIPTION Variable. Generally greyish-brown above, with paler stripe on cheek, and occasionally with darker dorsal stripe from top of head to upper back. Underparts, including legs and base of tail, pale yellowish-brown below. Tail darker distally, and occasionally with whitish tip. DISTRIBUTION North-eastern Qld, where fragmented colonies occur from around Mitchell River (north), Burdekin River (south) and Mungana (west). HABITAT AND HABITS Found in rocky areas, mainly in open forests and grassland, up to about 1,000m. Shelters during most of the day in caves and rock crevices, emerging late in the afternoon to feed on grasses and other plant material.

■ Kangaroos & Wallabies ■

Brush-tailed Rock-wallaby
■ *Petrogale penicillata* TL 1,010–1,285mm, including tail 500–700mm

DESCRIPTION Brown above, more greyish on shoulders and reddish on rump, with pale cheek-stripe and blackish patch near shoulder. Tail darkens towards brushed tip. Underparts paler, occasionally white on chest and feet blackish. **DISTRIBUTION** Eastern Australia, from south-eastern Qld, through eastern NSW and ACT, to north-eastern Vic, and reintroduced to Grampians in central western Vic. **HABITAT AND HABITS** Found in rocky areas in rainforests, forests and woodland. Nocturnal, feeding primarily on grasses supplemented with other plant material, including flowers, leaves, fruits, bark and fungi, and sheltering by day in rock crevices, on ledges and in caves. Lives in colonies of up to 30 individuals.

Proserpine Rock-wallaby
■ *Petrogale persephone* TL 1,120–1,320mm, including tail 520–670mm

DESCRIPTION Dark purplish-grey, with dark dorsal line from forehead to upper back, and blackish patch near armpit. Head has greyish forehead, and ears pale reddish-brown at bases and with paler outside edges. Tail dark above and pale below, normally with white tip. **DISTRIBUTION** Restricted to a number of subpopulations in small area of eastern Qld, including Clarke Range, Conway Range, Dryander, Gloucester Island and Hayman (introduced) Island. **HABITAT AND HABITS** Shelters in rocky areas in vine thickets and margins of rainforests on mainland, and *Acacia* woodland on Gloucester Island. Feeds on grasses, leaves, ferns and fungi, often foraging in more open savannah woodland, and occasionally residential gardens.

■ Kangaroos & Wallabies ■

Purple-necked Rock-wallaby
■ *Petrogale purpureicollis* TL 950–1,210mm, including tail 450–600mm

DESCRIPTION Pale brown above, more greyish on shoulders, becoming purple on neck. Longer dark brown guard hairs on lower back, and longitudinal dark brown dorsal stripe

from forehead to upper back. Tail darkens towards tip. DISTRIBUTION Discontinuous in western Qld, from Lawn Hill, east to around Winton. HABITAT AND HABITS Lives in rocky areas in woodland and spinifex grassland. Shelters by day in rocky crevices, under overhangs and in caves, emerging around dusk to feed on grasses and other plant material. Lives in colonies.

Eastern Short-eared Rock-wallaby
■ *Petrogale wilkinsi* TL 607–1,187mm, including tail 297–517mm

DESCRIPTION Dark grey-brown, washed with orange on legs and tail, with black dorsal stripe from crown to upper back, and black crescent, edged with white, behind forearms. Conspicuous silverish streaking on back and sides. Underparts greyish and feet black. Tail with blackish tip. DISTRIBUTION Northern and eastern NT to Qld border, and

English Company, Groote Eylandt, Sir Edward Pellow and Wessel Island Groups. HABITAT AND HABITS Lives in open, grassy woodland, where it favours rocky outcrops, cliffs and gorges. Mostly nocturnal, but may also feed during cooler days on grasses, herbs, seeds, fruits and other plant material, and shelters at other times in caves and similar places.

Kangaroos & Wallabies

Yellow-footed Rock-wallaby
■ *Petrogale xanthopus* TL 1,050–1,365mm, including tail 600–715mm

DESCRIPTION Pale brownish-grey above, with longitudinal blackish dorsal stripe from forehead to upper back, russet patch near armpit, and two-toned white-and-brown stripe on hip. Tail long and orange-brown, with dark brown rings. Underparts white. **DISTRIBUTION** Highly fragmented in former range. Occurs in south-western Qld, western NSW, and areas of southern and eastern SA. **HABITAT AND HABITS** Lives in rugged rocky areas. Shelters by day in rocky crevices and caves, although may be active diurnally in cooler weather, and feeds at night on grasses and herbs supplemented with seeds, fruits and flowers. Lives in small colonies, with some becoming quite large.

Rufous-bellied Pademelon
■ *Thylogale billardierii*
TL 970–1,200mm, including tail 320–415mm
(Tasmanian Pademelon)

DESCRIPTION Dark brown to greyish-brown above, flecked with paler brown, and pale brown below, tinged with reddish-brown on belly. Tail short (about 35 per cent of head-body) and thickened. **DISTRIBUTION** Formerly in southern Vic and south-eastern SA, but now only in Tas and larger adjacent islands. **HABITAT AND HABITS** Occupies wet forests, including rainforests, and wet gullies in sclerophyll forests. Shelters by day under dense vegetation. Forages at night in clearings within 100m of cover, on grasses, herbs and some flowers.

Kangaroos & Wallabies

Red-legged Pademelon
Thylogale stigmatica TL 690–1,010mm, including tail 300–470mm

DESCRIPTION Soft thick fur, greyish-brown above, more rufous on hindlegs, flanks, forearms and cheeks; pale brownish stripe on hip. Underparts off-white to greyish. Rainforest dwellers are darker than those living in vine thickets. Tail is short, about 75–85 per cent of head–body. **DISTRIBUTION** On ranges and slopes from Cape York in Qld to the Wyong area on the Central Coast of NSW. **HABITAT AND HABITS** Found in dense understorey and groundcover in rainforests, wet sclerophyll forests and vine thickets. Active late afternoon to early morning, but stays in dense vegetation during day and forages in more open areas at night. Feeds on native grasses and herbs in forest ecotone, seedlings and stem leaves, fungi, ferns, fruits and grasses. More nocturnal in open areas than similar Red-necked Pademelon (below).

Red-necked Pademelon
Thylogale thetis TL 620–1,110mm, including tail 300–510mm

DESCRIPTION Generally brownish above, more rufous on forehead, neck and shoulders, and flecked with grey on back, thighs and tail. Belly off white, becoming more white on chest and throat. Tail short, about 45 per cent of head-body. **DISTRIBUTION** Coast and range of eastern mainland, from around Wollongong NSW to near Gladstone Qld. **HABITAT AND HABITS** Inhabits rainforest fringes and wet sclerophyll forests. Occupies

dense vegetation by day, and feeds on grasses and leaves in adjacent open areas between dusk and dawn. Often occurs in same areas as Red-legged Pademelon (above), but more crepuscular in open areas.

KANGAROOS & WALLABIES

Spectacled Hare-wallaby
Lagorchestes conspicillatus TL 905–990mm, including tail 425–470mm

DESCRIPTION Greyish-brown above, with tips of hairs yellowish, and orange patch around eye that extends backwards slightly along cheek. Head has short, thickset snout, and orange patches at bases of ears. Underparts whitish to pale brown. **DISTRIBUTION** Patchily distributed through northern Australia, from Pilbara WA, through central and northern NT, to Rockhampton Qld. Also Barrow and Hermite (reintroduced) Islands. **HABITAT AND HABITS** Inhabits open woodland and hummock grassland. Mostly solitary, feeding at night on green grasses and legumes, and sleeping by day in dense vegetation, in spinifex tussocks or a short burrow.

Rufous Hare-wallaby
Lagorchestes hirsutus TL 555–695mm, including tail 245–305mm

DESCRIPTION Reddish-brown above, more greyish on individuals from WA, and paler below. Forearms paler than rest of body, and tail dark brown (proportionately smaller in WA). **DISTRIBUTION** Bernier, Dorre and Trimouille (introduced) Islands WA, and recently introduced into mainland conservation areas at Lorna Glen WA, Watarrka and Uluru Kata Tjuta National Parks NT, and Scotia Sanctuary, NSW. **HABITAT AND HABITS** Favours vegetated sandplains on islands. Shelters by day in short burrows or a scrape in the ground under low vegetation, and emerges at night to feed on succulent plant material.

▪ Kangaroos & Wallabies ▪

Western Grey Kangaroo
▪ *Macropus fuliginosus* TL 1,370–2,225mm, including tail 425–1,000mm

DESCRIPTION Fur shaggy. Brownish-grey above (more blackish on Kangaroo Island), slightly paler and more greyish below, with finely haired black ears and snout. Males have distinctive strong odour. **DISTRIBUTION** Southern Australia, from south-western WA (south of Shark Bay), through southern SA (including Kangaroo Island), western Vic and western NSW to inland of southern Qld. **HABITAT AND HABITS** Found in variety of wooded and grassland habitats. Mainly crepuscular and nocturnal, resting during heat of the day in shade, and emerging to forage on grasses, herbs and leaves of shrubs.

Male *Female and joey*

Eastern Grey Kangaroo
▪ *Macropus giganteus* TL 1,404–2,400mm, including tail 430–1,100mm

DESCRIPTION Grey to brownish-grey above, paler on face and hindlegs, and paler grey to whitish below. Head has moderately short, somewhat rounded ears, darker eye-ring and hairy snout. Feet and tip of tail black. **DISTRIBUTION** Eastern Australia, from around Cooktown Qld, through majority of NSW, ACT and Vic to south-eastern SA and eastern Tas (introduced to Maria Island). **HABITAT AND HABITS** Wide ranging in variety of wooded habitats and adjacent open areas, where it tends to favour wetter areas, but also occurs in semi-arid regions. Cathemeral, but rests during heat of day. Feeds on a variety of grasses, herbs and leaves.

Kangaroos & Wallabies

Agile Wallaby
Notamacropus agilis TL 1,180–1,700mm, including tail 587–840mm

DESCRIPTION Yellowish-brown above, with pale buff stripe on thigh, and whitish below. Head has rounded snout, indistinct pale brown cheek-stripe and short, dark brown, longitudinal stripe on forehead (not always present). **DISTRIBUTION** Tropical northern Australia (including offshore islands), from Kimberley region WA, through northern NT and coastal northern Qld to around Bundaberg in south. Also New Guinea. **HABITAT AND HABITS** Found in grassy woodland and adjacent grassland. Lives in social groups, grazing on grasses, fruits and sedges, mostly late in the afternoon and at night, and shelters during high temperatures in shade.

Black-striped Wallaby
Notamacropus dorsalis TL 1,070–1,650mm, including tail 540–830mm

DESCRIPTION Medium-sized wallaby, grey to brown above and rufous brown on upper back and neck, with paler sides. Black dorsal stripe from top of head to rump, along centre of back; white horizontal line on thigh. Face has white cheek stripes; nose, forepaws and toes are black. Underparts are much paler to white. Tail slightly shorter than body. Similar to Red-necked Wallaby (see p. 84), which is slightly larger and lacks black dorsal stripe. **DISTRIBUTION** Coastal and inland eastern Australia, from Chillagoe, southern Cape York Peninsula Qld, to New England and Far North Coast regions NSW. **HABITAT AND HABITS** Forests and woodland with a dense understorey. Can be observed in large groups, resting by day in dense vegetation and emerging at dusk to feed during the night in open areas on grasses and herbs.

■ Kangaroos & Wallabies ■

Tammar Wallaby
■ *Notamacropus eugenii* TL 850–1,130mm, including tail 330–450mm

DESCRIPTION Brown above, heavily streaked with silver-grey, and more reddish on flanks and legs (more so in males). Whitish stripe on cheek below eye, and pale stripe on thigh. Underparts paler grey-brown. **DISTRIBUTION** South-west mainland and East, Garden, Middle, North, North Twin Peaks and West Wallabi Islands WA. Kangaroo Island (extant), numerous smaller islands (introduced) and Innes National Park SA (reintroduced). **HABITAT AND HABITS** Occurs in coastal heaths, shrubland and forests with a dense understorey. Shelters in dense vegetation during most of the day, and feeds at night on grasses and some shrubs in more open areas.

Western Brush Wallaby
■ *Notamacropus irma* TL 1,500–1,850mm, including tail 600–950mm
(Kwoora)

DESCRIPTION Uniformly pale grey, with black longitudinal, raised dorsal stripe, and black paws and feet. Head has white cheek, facial stripe and two-toned black-and-white ears. **DISTRIBUTION** South-western WA from around Cape Arid to just north of Kalbarri National Park. **HABITAT AND HABITS** Occurs in variety of open wooded habitats and open grassland. Cathemeral, and more diurnal than other similar species in range, grazing on plant material and sheltering at other times.

◾ Kangaroos & Wallabies ◾

Parma Wallaby
◾ *Notamacropus parma* TL 855–1,075mm, including tail 405–540mm

DESCRIPTION Greyish-brown above, with darker longitudinal dorsal stripe and white cheek-stripe. Underparts whitish, paler on chest and throat, and tail often tipped with white. When hopping, tail is usually in a horizontal position to the ground. **DISTRIBUTION** Ranges and slopes of eastern NSW, from Qld border, south to around Gosford. **HABITAT AND HABITS** Moist eucalypt forest with thick, shrubby understorey, often with nearby grassy areas, rainforest margins and occasionally drier eucalypt forest. Feeds at night on grasses and herbs in more open eucalypt forest and edges of nearby grassy areas. Uses runways to move through the thickly vegetated understorey. Shelters during day in dense cover.

Whiptail Wallaby
◾ *Notamacropus parryi* TL 1,400–1,970mm, including tail 728–1,045mm
(Pretty-face Wallaby)

DESCRIPTION Colour variable, brownish above in summer and pale grey in winter, with pale stripe on hip, extending along black-tipped tail. Head has slender snout, and is blackish to brownish on top, bordered below by broad white cheek-stripe, and ears are dark brown at bases and tips, with white in between. **DISTRIBUTION** Eastern Australia, from around Cooktown Qld to around Grafton NSW. **HABITAT AND HABITS** Favours open forests with a grassy understorey and generally in hilly country. Feeds during the day or night on grasses, ferns and herbaceous plants, often in small to large groups. Rests in shade during heat of the day. Generally does not need to drink (except during droughts), getting moisture from plants.

◾ Kangaroos & Wallabies ◾

Red-necked Wallaby ◾ *Notamacropus rufogriseus* TL 1,282–1,785mm, including tail 623–876mm
(Bennett's Wallaby)

DESCRIPTION Grey-brown above, heavily flecked with white, and washed with reddish on head, shoulders, arms, upper back and base of tail. Underparts pale grey to whitish. Face has obscure cheek-stripe and blackish longitudinal line on forehead. Darker and more brownish in Tas. **DISTRIBUTION** South-eastern mainland, from south-eastern Qld, through eastern NSW, ACT and south Vic, to far south-eastern SA. Also on Tas and major islands (including Bass Strait). **HABITAT AND HABITS** Found in open forests and woodland, and adjacent heathland. Mostly solitary, although may be seen in groups of up to about 30, feeding from late afternoon to dawn on grasses and herbs; rests at other times in dense vegetation.

Tasmanian albino form

Bridled Nailtail Wallaby
◾ *Onychogalea fraenata* TL 790–1,240mm, including tail 360–540mm
(Merrin)

DESCRIPTION Dark yellowish-grey above, washed with rufous on neck, and with distinct white markings from middle of neck and behind armpits to belly. White stripe from nose to rear of eye and along base of cheek. **DISTRIBUTION** Restricted areas in eastern Australia, naturally occurring in Taunton National Park, eastern Qld, and introduced to Avocet Nature Reserve and Idalia National Park, central Qld, and Scotia Sanctuary, western NSW. **HABITAT AND HABITS** Lives in open grassy woodland, dense *Acacia* shrubland and adjacent open grassland. Mostly solitary, sheltering in dense vegetation or in hollow logs, and emerging in the late afternoon to feed on herbs and grasses.

KANGAROOS & WALLABIES

Northern Nailtail Wallaby
■ *Onychogalea unguifera* TL 1,090–1,440mm, including tail 600–740mm

DESCRIPTION Yellowish-brown above, washed with pale orange-buff (particularly on flanks), cream hip-strip, and blackish longitudinal dorsal stripe from neck to base of tail. Head paler, with cream cheek-stripe and around eye. Tail long, whitish in middle and with blackish tufted terminal third and hardened, flattened 'nail' at tip. Underparts and feet white. **DISTRIBUTION** Semi-arid tropical north, from southern Kimberley WA, through central and northern NT (including Groote Eylandt), to western Cape York Peninsula Qld. **HABITAT AND HABITS** Occurs in open, grassy woodland, shrubland and grassland, where it is generally solitary, feeding mainly on leaves of succulent herbaceous plants, young grass shoots and fruits. Mostly crepuscular and nocturnal, resting at other times under dense shrubs.

Antilopine Wallaby
■ *Osphranter antilopinus* TL 1,455–2,100mm, including tail 680–890mm

DESCRIPTION Males pale reddish-brown above, with black paws and feet, and large black nose. Females grey or reddish-brown, more greyish on head and shoulders, with white edges to ears, and slightly smaller black nose than males. Underparts off white to grey. Similar to smaller Agile Wallaby (see p. 81) but lacks facial stripe. **DISTRIBUTION** Tropical northern Australia, from Kimberley region WA, through northern NT, to northern Qld (absent from tip of Cape York Peninsula). **HABITAT AND HABITS** Found in open grassy woodland. Can occur in moderately large groups. Feeds from dusk to dawn on grasses in more open areas, and may be active by day in cooler weather, but rests during hotter days in shade under trees or rocky overhangs.

■ KANGAROOS & WALLABIES ■

Black Wallaroo
■ *Osphranter bernardus* TL 1,100–1,400mm, including tail 545–640mm

DESCRIPTION Males blackish-brown to blackish above, with black paws, feet and tip of tail. Females grey to greyish-brown, with dark brown tail-tip. **DISTRIBUTION**

Escarpment and plateau of western Arnhem Land NT. **HABITAT AND HABITS** Occurs in seasonally wet rainforests, open forests, woodland, heathland and spinifex grassland in areas of rocky sandstone. Mostly nocturnal, resting by day in rocky caves and crevices, and emerging to feed on a range of plant material, including grasses, fruits, seeds, flowers and tubers.

Common Wallaroo
■ *Osphranter robustus* TL 1,550–2,000mm, including tail 750–900mm
(Euro)

DESCRIPTION Heavily built (especially males), with shaggy fur and a characteristic hunched stance. Dark grey, yellowish-brown to reddish-brown or paler grey above, depending on subspecies, and generally paler below. Males darker than females.
DISTRIBUTION Throughout most of Australia and Barrow Island. Absent from far south (including Tas) and western Cape York Peninsula Qld. **HABITAT AND HABITS** Normally occurs in small groups of up to 5 in hilly country and rocky ranges in variety of wooded and grassland habitats. Feeds mainly on grasses and herbaceous plants (principally spinifex *Triodia* seeds on Barrow Island).

Female   *Male*

KANGAROOS & WALLABIES

Red Kangaroo
■ *Osphranter rufus* TL 1,645–2,400mm, including tail 645–1,000mm

(Marloo)

DESCRIPTION Largest marsupial. Adult males reddish-brown above, and females blue-grey, with squarish snout, and white stripe between mouth and ear. Underparts white in both sexes, but males may be stained on chest.

DISTRIBUTION Throughout arid and semi-arid inland Australia in all states except Tas. Reaches coast in north-western WA and on Nullarbor Plain.

HABITAT AND HABITS Inhabits grassland, shrubland and woodland. Mainly active from dusk to dawn, but may be sighted at any time of the day. Rests in shade during the heat of the day. Feeds on grasses and herbs, and must drink regularly.

Female and joey   *Male*

Swamp Wallaby
■ *Wallabia bicolor* TL 1,105–1,710mm, including tail 640–862mm

DESCRIPTION Dark reddish-brown to black above, with light pale yellowish-brown cheek-stripe (more prominent in northern individuals), light brown to rufous orange on chest and pale yellowish to orange-brown below. Extremities, such as tail, fore-arms and legs, show a darker, almost black colouring, and tail, occasionally with white tip.

DISTRIBUTION Broadly distributed within coast, slopes and ranges of eastern and south-eastern Australia (including Fraser Island), from northern Cape York Peninsula Qld, through eastern NSW, ACT and southern Vic, and possibly into far south-eastern SA.

HABITAT AND HABITS Forests, woodland and heaths with a dense understorey. Shelters during most of the day in dense vegetation, often in moist areas, and emerges at night to feed on grasses, shrubs, ferns and seedlings. When disturbed, bounds away with head low and tail held straight.

▪ Kangaroos & Wallabies ▪

Quokka ▪ *Setonix brachyurus* TL 680–850mm, including tail 245–310mm
(Ban-gup)

DESCRIPTION Uniformly brown, heavily flecked with grey and tinged with reddish-brown. Short, sparsely furred tail (about 60 per cent head-body length). Head has angled snout, short, rounded ears and naked, greyish-black nose. **DISTRIBUTION** South-western WA, including Bald and Rottnest Islands, although subpopulation area is small and heavily fragmented. **HABITAT AND HABITS** Found in forests, woodland, heathland, swamps and wetlands. Can be active day or night, feeding on variety of plant matter, but eats more succulent plants when surface water is scarce, and shelters at other times under vegetation.

Banded Hare-wallaby
▪ *Lagostrophus fasciatus* TL 750–850mm, including tail 350–400mm
(Maning)

DESCRIPTION Fur shaggy. Dark grey above, heavily streaked with silver, and with series of darker transverse bands on lower back and rump. Underparts pale whitish-grey. **DISTRIBUTION** Shark Bay WA. Naturally occurring on Bernier and Dorre Islands, and introduced to Faure Island in 2004–2012. **HABITAT AND HABITS** Occurs in sandplains and dunes. Shelters by day under dense vegetation, emerging to feed in open areas at night on variety of plant material, including grasses and shrubs.

DUGONG/RATS & MICE

DUGONGIDAE, DUGONG

Dugong ■ *Dugong dugon* TL 2.5–4.5m

DESCRIPTION Large, smooth body, generally grey to greyish-brown, paler and washed with pink below, with small flippers and large, horizontally flattened, triangular tail fluke. Head relatively small, with downwards-pointing mouth, surrounded by long bristles. No dorsal fin. **DISTRIBUTION** Northern Australia in shallow tropical seas and estuaries, from around Shark Bay WA to Moreton Bay Qld. **HABITAT AND HABITS** Grazes communally on seagrass beds in channels and sheltered estuaries during high tides (generally at 3–10m), but also eats marine algae and aquatic invertebrates, and rests at low tide in deeper offshore water. Female becomes sexually mature after about 7 years of age. Several males jostle for the chance to mate with a single female.

MURIDAE, RATS & MICE

Water Rat ■ *Hydromys chrysogaster* TL 460–665mm, including tail 230–320mm

DESCRIPTION Dark grey to blackish above, with moderately long, broad, hairy tail, usually with white tip. Head dorsally flattened, with small, rounded ears and blunt nose; nostrils face upwards high on snout to assist breathing while swimming. Underparts yellowish-orange or whitish. Partially webbed hindfeet. **DISTRIBUTION** Widespread on mainland and offshore islands in Australia, including south-western and northern WA, northern NT, northern and eastern SA, Qld, NSW, ACT, Vic and Tas. **HABITAT AND HABITS** Found near fresh water in rainforests, sclerophyll forests and woodland, and coastal saline waters. Cathemeral, foraging in streams and swamps, and on land for crustaceans, fish and large insects. Shelters at other times in underground tunnel.

RATS & MICE

Forrest's Mouse
Leggadina forresti TL 115–170mm, including tail 45–70mm

DESCRIPTION Heavily built. Generally yellowish-brown to greyish-brown above, flecked with black, and white below. Tail shorter than head-body, greyish and mostly naked. Ears small and rounded, with whitish patches around base. **DISTRIBUTION** Arid inland

Australia, including central and southern NT, eastern WA, northern SA, far north-western NSW and south-western Qld. **HABITATS AND HABITS** Mainly inhabits low shrubland and tussock grassland, but also found in woodland and sandy deserts. Shelters during the day in shallow underground burrow lined with grass, and feeds at night on seeds, insects and succulent parts of plants. Like other arid specialists, probably gains all its moisture from food.

Northern Short-tailed Mouse
Leggadina lakedownensis TL 100–120mm, including tail 40–45mm
(Kerekenga)

DESCRIPTION Grey to greyish-brown above and white below, including feet. Upper incisors angled forwards, which identifies it from Forrest's Mouse (see above).

DISTRIBUTION Northern Australia, from Pilbara region, Thevenard and Serrurier (introduced) Islands WA, through northern NT to northern Qld. **HABITAT AND HABITS** Mainly found in grassland, but also woodland, shrubland and rocky outcrops. Nocturnal, sleeping by day in shallow burrows, and feeding at night on seeds, grass and small invertebrates.

RATS & MICE

Greater Stick-nest Rat
■ *Leporillus conditor* TL 335–440mm, including tail 145–180mm
(Wopilkara)

DESCRIPTION Greyish-brown above with yellowish wash, and creamish-white below (more brownish on underside of tail). Ears large, broad and rounded, and snout somewhat blunt. **DISTRIBUTION** Naturally occurring on East and West Franklin Islands SA, introduced to Reevesby and Saint Peter Islands SA, Salutation and Faure Islands WA, Roxby Downs Arid Recovery Reserve SA and Scotia Sanctuary NSW. **HABITAT AND HABITS** Occupies dense shrubland. Shelters by day in family groups of up to about 20 within a large nest of sticks (up to 1m high), and feeds at night primarily on leaves and fruits of succulents, supplemented with grasses.

Broad-toothed Rat
■ *Mastacomys fuscus* TL 240–305mm, including tail 100–130mm
(Tooarrana)

DESCRIPTION Fur long and dense, and pale to dark brown above, flecked with russet and yellow, and pale brownish-grey below. Head broad, with large cheeks and small, rounded ears. Tail short and sparsely furred. **DISTRIBUTION** Isolated subpopulations in south-eastern mainland, including Barrington Tops NSW, Brindabella Range ACT, Cann River, Dandenong Ranges and Wilsons Promontory Vic, and western Tas. **HABITAT AND HABITS** Terrestrial in broad range of vegetated habitats in coastal to alpine regions. During winter, sleeps by day in small groups in grassy nest, and during summer, in underground burrows, emerging to feed mainly on grasses.

▪ Rats & Mice ▪

Grassland Melomys
▪ *Melomys burtoni* TL 180–340mm, including tail 90–170mm (smallest on east coast)

DESCRIPTION Highly variable. Generally dark grey to reddish-brown, occasionally more pale brown or pale orange on sides, and with thin, scaly tail, greyish on top and pinkish below. Underparts, including feet, greyish. **DISTRIBUTION** Coastal northern and eastern Australia (including offshore islands), including northern Kimberley WA and northern NT, and from northern Cape York Peninsula Qld to around Gosford NSW. **HABITAT AND HABITS** Prefers tall grassland, but also occurs in grassy areas in rainforests, forests, woodland and mangroves. Mainly terrestrial, but also an agile climber, feeding at night on grasses, fruits and insects. Shelters by day in nests in tree hollows, shrubs, hollow logs or burrows. Thrives in cultivated sugarcane fields, becoming a major pest.

Cape York Melomys
▪ *Melomys capensis* TL 240–334mm, including tail 129–172mm

DESCRIPTION Highly variable. Generally pale greyish-brown washed with orange, and with thin, scaly, sparsely furred tail (longer than head-body), brownish on top and pinkish-grey below. Underparts, including feet, white to creamish-fawn. **DISTRIBUTION** Northern Cape York Peninsula, Qld. **HABITAT AND HABITS** Found in rainforests, wet sclerophyll forests and woodland. Semiarboreal, feeding at night on leaves, fruits and seeds. Also enters houses adjacent to natural habitat. Shelters by day in nests of dried leaves in tree hollows.

RATS & MICE

Fawn-footed Melomys

Melomys cervinipes TL 215–400mm, including tail 115–200mm

DESCRIPTION Highly variable. Brown to brownish grey dorsal surface, with a greyish face. Underparts white to creamish fawn including feet. Has a mosaic pattern of scales on the tail compared to Rattus spp., which have a ringed pattern of scales. Often referred to as 'Mosaic-tailed Rat'. Tail can be slightly prehensile and is sparsely furred, grey-brown to blackish and generally similar or slightly longer in length to head–body. **DISTRIBUTION** Eastern Australia (including Fraser Island), from just north of Cooktown in north-east Qld south to the Central Coast NSW. **HABITAT AND HABITS** Occurs in closed tropical forests (including rainforests, vine thickets and wooded swamps) in much of Qld. Within southern limits in NSW, found in wet sclerophyll forests, coastal woodlands and mangrove forests. Mainly nocturnal and arboreal, feeding at night on leaves and fruits, and nesting during the day in spherical nests of leaves and grasses in tree canopy. Breeding occurs mainly spring to summer but can breed anytime other than winter producing young more than once during the year.

Black-footed Tree-rat

Mesembriomys gouldii TL 585–720mm, including tail 310–410mm (Djintamoonga)

DESCRIPTION Fur long, grey to blackish above, darker on rump, and heavily streaked with silver and yellowish-brown. Tail furred and blackish, tipped with white, and longer than head-body length. Underparts pale greyish or white. Ears and feet blackish. **DISTRIBUTION** Northern Australia, from northern Kimberley region NT, northern NT, including Melville Island, and north-eastern Qld. **HABITAT AND HABITS** Lowland forests and woodland with dense understorey. Sleeps by day mostly in tree hollows, emerging to forage both on the ground and in trees for fruits and seeds, supplemented with flowers, grasses and invertebrates.

RATS & MICE

Golden-backed Tree-rat
■ *Mesembriomys macrurus* TL 480–605mm, including tail 290–360mm

DESCRIPTION Generally greyish, heavily washed with dark yellowish-brown along mid-dorsal area from crown to rump, and whitish below. Tail long (about 150 per cent of head-body), greyish at base and with white brushed tip.

DISTRIBUTION North-western Australia, from northern Kimberley region (including several offshore islands) WA to northern NT, although no recent records from NT. **HABITAT AND HABITS** Inhabits rainforests, woodland and rocky sandstone areas. Crepuscular and nocturnal, feeding mostly in trees, but also on the ground in less vegetated areas, mainly on fruits, leaves and invertebrates, including termites, but also on grasses. Sleeps at other times in tree hollows and rocky crevices.

House Mouse
■ *Mus musculus* TL 135–190mm, including tail 75–90mm

DESCRIPTION Pale brown to grey, with whitish or pale grey-brown belly. Ears large and rounded. Tail mostly naked, with circular annulations (rows) of scales clearly visible. Notched incisor and musty odour can distinguish it from superficially similar species, including smaller *Pseudomys spp*. Female has 4 teats. **DISTRIBUTION** Introduced, most likely from the Mediterranean. Found in suitable habitats throughout Australia, often in close association with humans.

HABITAT AND HABITS Mostly nocturnal, in a range of habitats, both urban and natural. Highly adaptable, which has enabled it to thrive. Usually lives in cracks or within a complex network of underground tunnels, often constructed under cover, such as debris. Food typically consists of seeds and other plant material, fungi, small invertebrates and human food.

RATS & MICE

Spinifex Hopping-mouse
Notomys alexis TL 215–260mm, including tail 120–150mm
(Tarrkawarra)

DESCRIPTION Pale brown above, washed with reddish-orange, with black guard hairs, and greyish face and snout. Underparts white to greyish-white, with small, naked throat-patch. Tail long (about 125 per cent of head-body), with fine tufted tip of whitish-grey hairs. Rear feet elongated. Similar to **Dusky Hopping-mouse** *N. fuscus*, which has slightly longer and darker, more brush-tipped tail. **DISTRIBUTION** Arid central Australia, from central west coast WA (except Pilbara), through southern NT and northern SA, to western Qld. **HABITAT AND HABITS** Occupies spinifex sandy deserts and arid shrubland. Shelters by day in deep communal burrows lined with leaves and small twigs, emerging at dusk to forage on the ground for plant material, mainly seeds, and small invertebrates.

Fawn Hopping-mouse
Notomys cervinus TL 215–280mm, including tail 120–160mm
(Ooarri)

DESCRIPTION Buff to greyish above, including top of tail, with large ears and rounded snout; white below. Tail longer than head–body with darker brushed tip. Similar to other species of hopping-mouse in range, but lacks throat-pouch. **DISTRIBUTION** Possibly restricted to channel country of south-western Qld and north-eastern SA, although population numbers capable of rapid increases following extended high rainfall periods, and records exist from western NSW, southern NT, central Qld, and further west and south in SA. **HABITAT AND HABITS** Lives in stony (gibber) plains and claypans, where it sleeps by day in communal underground burrows and feeds at night on seeds, supplemented with grass and arthropods.

▪ Rats & Mice ▪

Mitchell's Hopping-mouse
▪ *Notomys mitchellii* TL 240–280mm, including tail 150–155mm
(Pankot)

DESCRIPTION Grey-brown above, tinged with yellowish-orange, and heavily flecked with darker and paler grey. Tail dark pinkish-brown, long and with darker brownish brush tip. Underparts white to greyish-white. Larger and more greyish than similar Fawn Hopping-mouse (see p. 95). **DISTRIBUTION** From south-western WA, through southern SA, to south-western NSW and north-western Vic. **HABITAT AND HABITS** Found in mallee shrubland and dense heathland, mainly on sandy soils. Nocturnal, sheltering by day in deep burrows, and emerging to feed mainly on seeds, supplemented with insects, and more succulent parts of plants in drier weather.

Ash-grey Mouse
▪ *Pseudomys albocinereus* TL 148–210mm, including tail 85–110mm
(Noodji)

DESCRIPTION Silver-grey above, occasionally tinged with pale brown, and white below. Feet and tail pink, with tail longer than head-body length, and sometimes with darker

base. Similar to slightly larger and longer tailed **Western Mouse** *P. occidentalis*. **DISTRIBUTION** South-western WA, although absent from far south-west, from Shark Bay (including Bernier, Dorre and Dirk Hartog Islands) to Israelite Bay. **HABITAT AND HABITS** Occupies heaths and shrubland on sandplains, and coastal sandplains on islands. Feeds at night primarily on seeds and green plant material, supplemented in drier summer months with insects. Sleeps by day in communal burrows, sometimes only consisting of family members.

RATS & MICE

Plains Mouse
Pseudomys australis TL 180–260mm, including tail 80–120mm
(Palyoora)

DESCRIPTION Grey to yellowish-grey above (including top of tail), with darker grey-brown guard hairs, and tail shorter than head-body length. Underparts and underside of tail white to creamish-white.
DISTRIBUTION Recent records restricted to central and northern SA, southern NT and south-western Qld, although range likely to be larger, particularly following periods of extensive rainfall. **HABITAT AND HABITS** Found in gibber plains and cracking claypans, with adequate shrubby groundcover. Nocturnal, foraging on the ground for seeds, supplemented with leaves and grasses, and some invertebrates. Shelters by day either in small groups or solitarily in shallow burrows or soil crevices.

Delicate Mouse
Pseudomys delicatulus TL 111–155mm, including tail 55–84mm
(Molinipi)

DESCRIPTION Yellowish-brown or orange-brown to pale brownish-grey above, with black guard hairs, and white or cream below (including underside of tail). Nose and feet pink, and top of tail pale brownish. **DISTRIBUTION** Tropical northern Australia (including offshore islands), from northern Pilbara region WA, through northern NT and northern Qld, to Fraser Island in south. **HABITAT AND HABITS** Occurs in sparsely vegetated grassland and coastal sand dunes with softer soils. Sleeps by day in underground burrows or, occasionally, in termite mounds, and forages at night on plant materials supplemented with insects.

▪ Rats & Mice ▪

Desert Mouse ▪ *Pseudomys desertor* TL 140–210mm, including tail 67–105mm
(Wildjin)

DESCRIPTION Fur shaggy. Rich reddish-brown above, with conspicuous longer, darker guard hairs, and pale brownish-orange ring of naked skin around eye. Underparts pale grey, becoming more off white on underside of tail. Similar to larger Western Chestnut Mouse (see p. 101); ranges overlap in western Tanami Desert. **DISTRIBUTION** Arid central Australia, including western WA, southern NT, western Qld and northern SA. Also Bernier Island WA. **HABITAT AND HABITS** Lives in mature hummock grassland and shrubland on sandy plains or rocky slopes. Shelters, usually solitarily, in a shallow burrow and forages partly during the day, but mainly at night, for seeds and succulent parts of plant stems.

Smoky Mouse ▪ *Pseudomys fumeus* TL 195–285mm, including tail 105–150mm
(Konoom)

DESCRIPTION Pale grey to black above (darkest in west), often tinged with bluish, and blackish around eye and on top of snout. Tail dark greyish-brown on top and pinkish below. Underparts greyish-white. **DISTRIBUTION** Patchily distributed in coastal slopes and ranges of south-eastern mainland, from southern NSW to western Vic. **HABITAT AND HABITS** Occurs in various habitats that have been subject to some – but not too frequent – fires, including dry and wet forests and woodland with a dense heath understorey, and heathland. Shelters by day, and feeds at night on plant material, including seeds, berries and flowers, subterranean fungi and invertebrates.

= RATS & MICE =

Eastern Chestnut Mouse
■ *Pseudomys gracilicaudatus* TL 185–265mm, including tail 80–120mm (Karrooka)

DESCRIPTION Brown to grizzled chestnut on dorsal surface and greyish on ventral surface and on flanks, flecked silvery-grey, with obscure pale eye-ring. Greyish fur below, feet with long white hairs extending beyond claws, distinguishing it from Swamp Rat (see p. 107), which has much darker greyish feet. Eyes and ears are small. Tail sparsely hairy and shorter than head–body. DISTRIBUTION Disjunct populations along coast and ranges of eastern Australia, from Cooktown in north Qld, west to Emerald and down to Jervis Bay on south coast NSW. HABITAT AND HABITS Heathland and open grassy woodland that is regularly burnt. Generally more common in dense wet heath and swamps. In tropics to sub-tropics more common in open grassy woodland. Sleeps during most of the day in nests in dense vegetation above ground or in hollows; forages from dusk to dawn within runways in thick vegetation for seeds, plant stems, fungi and insects. More numerous in areas affected by fire.

Sandy Inland Mouse
■ *Pseudomys hermannsburgensis* TL 135–175mm, including tail 70–90mm

DESCRIPTION Yellowish-brown to brown above, flecked with blackish-brown and silvery-grey, and white below. Tail slightly longer than head-body, pinkish-brown and sparsely haired. Distinguished from House Mouse (see p. 94) by longer tail, larger ears, absence of notch on upper incisors and lack of musty odour. DISTRIBUTION Arid and semi-arid inland, reaching coast and offshore islands of central western WA, and extending through central and southern NT, central and northern SA, western Qld and north-western NSW. HABITAT AND HABITS Found in hummock grassland and scrubland on softer soils. Shelters in small social groups during the day in deep underground burrow, and forages at night for seeds, succulent shoots, underground tubers and insects.

▪ Rats & Mice ▪

Long-tailed Mouse
▪ *Pseudomys higginsi* TL 260–350mm, including tail 145–200mm
(Looringa)

DESCRIPTION Largest member of genus. Fur moderately long and soft. Dark grey above and paler grey below (more whitish on feet), with long, bicoloured tail. Black markings on

sides of snout and around eye. DISTRIBUTION Tas. HABITAT AND HABITS Found in rainforests with high annual rainfall and alpine heaths. Omnivorous and mostly nocturnal, foraging in runways in dense leaf litter, or in tunnels under mosses, for variety of plant material, fungi and arthropods. Shelters at other times in nest placed in hollow log or rocky crevice.

Central Pebble-mouse
▪ *Pseudomys johnsoni* TL 150–170mm, including tail 75–95mm
(Ilyema)

DESCRIPTION Greyish-brown above, with longer black guard hairs, and more greyish on head. White underparts, including feet and extending up mouth. Very similar to generally smaller Delicate Mouse (see p. 97) and Sandy Inland Mouse (see p. 99), which

has shorter ears and longer tail. DISTRIBUTION Patchy within range in semi-arid northern Australia, from western Qld, through central and north-western NT, to northern Kimberley region WA. HABITAT AND HABITS Marked preference for rocky, pebble-covered hills in eucalypt woodland. Shelters by day in social groups in underground burrows, with small pebble mounds placed around entrance. Forages on the ground at night.

RATS & MICE

Western Chestnut Mouse

■ *Pseudomys nanus* TL 150–260mm, including tail 70–120mm (Moolpoo)

DESCRIPTION Pale orange-brown above, with longer dark brown guard hairs, and fur below whitish with grey base. Tail dark brown above and white below, and shorter than head-body. Pale chestnut eye-ring. Similar to smaller and darker Desert Mouse (see p. 98). DISTRIBUTION Tropical northern Australia, from southern Kimberley region WA, through central and northern NT, to far north-western Qld. Also Barrow Island WA. HABITAT AND HABITS Found in low eucalypt woodland with dense tussock grass understorey, and often along watercourses. Shelters in grass nest by day, and feeds mainly on grass stems at night.

New Holland Mouse

■ *Pseudomys novaehollandiae* TL 145–195mm, including tail 80–107mm (Pookila)

DESCRIPTION Grey-brown above, streaked with silverish and paler brown hairs. Underparts generally white. Tail dusky brown on top and white on underside, longer than head–body. Feet white on upper surface. Similar in size to House Mouse (see p. 94), although has larger ears and eyes. Also lacks notch on upper incisors and does not have distinctive musty 'mouse' odour. DISTRIBUTION Fragmented populations in Tas, Vic, NSW and Qld. Coastal and near coastal south-eastern mainland, eastern Tas and Flinders, and Three Hummock Islands in Bass Strait. Patchy on mainland from far south Qld, through eastern NSW, to southern Vic. HABITAT AND HABITS Open heathland, woodland, open forests with a heathland understorey and vegetated sand dunes. Occupies variety of complex habitats, with seemingly a preference for areas in midstages of growth following fire. Shelters communally by day in burrows, emerging at night to feed on plant material, including seeds, leaves and roots, fungi and invertebrates.

RATS & MICE

Hastings River Mouse
- *Pseudomys oralis* TL 240–325mm, including tail 120–160mm
(Koontoo)

DESCRIPTION Generally brownish-grey, streaked with silverish and blackish-brown. Tail furry with long hairs laying along the tail, dark brown above and white below. Underparts pale brown to greyish-white, feet white with long hairs extending past claws. Eyes distinctly bulging with a black eye ring in adults. DISTRIBUTION Scattered colonies along Great Dividing Range from south of Muswellbrook in Hunter Valley NSW, north-east NSW, Warwick, south-east Qld, and across towards Tweed Heads in NSW. HABITAT AND HABITS Dry sclerophyll forests, woodland, and within the ecotone of wet sclerophyll forests with dense groundcover and patches of loose rock. Shelters by day in fallen logs, hollow roots, tree crevices and piles of rocks. Nocturnal, feeding mainly on seeds and fruits during summer, and leaves and stems during winter when seeds are scarce.

Eastern Pebble-mouse
- *Pseudomys patrius* TL 120–160mm, including tail 63–80mm

DESCRIPTION Upperparts yellow-brown with numerous long, black guard hairs. Underparts whitish, extending to buff flanks, sides of mouth and just below eye. Feet pinkish above with short white hairs, and tail dark pinkish, sparsely haired with obvious rings of scales. Head long and somewhat flattened. Similar to Delicate Mouse (see p. 97), but head and muzzle longer and incisors broader. DISTRIBUTION Eastern Qld, between Paluma and Gympie. HABITAT AND HABITS Dry, hilly, rocky landscapes within grassy woodland, east of the Great Dividing Range. Occupies areas with sufficient pebbles, although degraded sites are avoided. Constructs conical mounds of pebbles containing entrances to its numerous underground burrows, generally at base of trees or amongst larger rocks.

Rats & Mice

Brown Rat

■ *Rattus norvegicus* TL 430mm, including tail 180mm

DESCRIPTION Extremely varied from brown to grey brown. Lighter coloured underneath. Ears and tail are shorter than those of Black Rat (below). **DISTRIBUTION** Introduced, most likely with arrival of Europeans in the 18th century. Usually found around ports, human habitation, with strong holds around cities. Origin is thought to be from Caspian region of eastern Europe. **HABITAT AND HABITS** Nocturnal, living on ground and beneath debris. Will dig burrow complexes. Does not usually climb. Feeds on animals including insects, mice, birds, eggs and lizards. Will also eat grain, seed and scraps. Unpleasant disposition, readily biting if provoked.

Peter Rowland/Kape Images

Black Rat

■ *Rattus rattus* TL 420 mm, including tail 200mm

DESCRIPTION Extremely varied from white to black but usually slate grey. Under body is white to cream. Ears and tail are much longer than those of Brown Rat (above). Introduced **Pacific Rat** *R. exulans*, found on islands of Australia's north, has smaller ears that cover the eyes if pulled forward. **DISTRIBUTION** Introduced, most likely with the arrival of Europeans in 18th century. Usually found around human habitation, with strongholds around cities but is found well away from human settlement. Origin is thought to be the Middle East. **HABITAT AND HABITS** Nocturnal, living above ground. Often forms nests in roofs and trees. Feeds on animals including insects, mice, birds, eggs and lizards. Will also eat grain, seed and scraps. Not as aggressive as Brown Rat but will still defend itself.

Angus McNab

= RATS & MICE =

Giant White-tailed Rat
■ *Uromys caudimaculatus* TL 598–744mm, including tail 323–362mm
(Mati)

DESCRIPTION Large rodent. Greyish-brown above with darker guard hairs, and paler, more greyish flanks. Snout partially naked, underparts creamish-white and feet pinkish-white. Tail naked, with blackish on base, terminal third white, and varying amounts of dark grey in between. **DISTRIBUTION** Coastal and near coastal north-eastern Qld, from Townsville to northern Cape York Peninsula. **HABITAT AND HABITS** Lives in rainforests, closed forests, wetter open forests and mangroves. Shelters in burrows or tree hollows by day. Omnivorous, feeding both in trees and on the ground, and eats variety of plant material, including fruits and seeds, fungi, insects, birds' eggs, crustaceans and vertebrates. Some food, particularly seeds, is cached (buried) for later consumption.

Common Rock-rat
■ *Zyzomys argurus* TL 164–246mm, including tail 90–125mm
(Dory)

DESCRIPTION Slender, small rodent. Golden-brown or dark grey-brown to blackish above, with darker guard hairs, and white to pale grey below. Tail usually longer than head-body, thick and lightly haired towards tip. When skin on tail is damaged, the tail rots away, often leaving a stump. **DISTRIBUTION** Patchy, throughout northern Australia

and several offshore islands, including Pilbara WA, Kimberley WA, northern NT and much of Qld. **HABITAT AND HABITS** Occupies variety of habitats, but always associated with rocky outcrops with adequate crevices and loose boulders, and generally with adjacent dense vegetation. Forages at night for plant stems, seeds, fungi and some insects, and sleeps by day in rocky crevices.

RATS & MICE

Central Rock-rat
■ *Zyzomys pedunculatus* TL 220–280mm, including tail 116–128mm
(Antina)

DESCRIPTION Yellowish-brown above, more pale brown on sides and whitish-cream below. Tail pale brown, thickened and hairy, with tufted tip. Tops of forefeet and hindfeet pale buff. **DISTRIBUTION** West MacDonnell Ranges in southern NT. **HABITAT AND HABITS** Core habitat consists of rocky quartzite ridges with patches of spinifex and some trees and shrubs, but also moves into adjacent rocky areas, open woodland and grassland. Forages terrestrially at night for plant material, mainly seeds and stems, supplemented with some invertebrates.

Kimberley Rock-rat
■ *Zyzomys woodwardi* TL 200–305mm, including tail 95–135mm
(Djookooropa)

DESCRIPTION Cinnamon-brown above, darker on head, and flecked with dark brown. Greyish below, with white tips to hairs giving more whitish appearance. Tail short, moderately furred and swollen at base. Tail skin very fragile, and tail often stunted and damaged from fighting.
DISTRIBUTION Higher rainfall areas of northern Kimberley region WA, including several offshore islands. **HABITAT AND HABITS** Found in rocky boulder outcrops on slopes in rainforests and open woodland, where it is mostly nocturnal. Feeds mainly on larger seeds of rainforest trees, supplemented with smaller grass seeds and fibrous plant material.

Fragile tail, often lost or damaged during fighting

RATS & MICE

Dusky Rat ■ *Rattus colletti* TL 215–360mm, including tail 95–150mm
(Marrawata; Mulbu)

DESCRIPTION Dark brown to blackish above, with numerous coarse, greyish guard hairs, and becoming more yellowish on flanks and yellowish-grey on underparts. Tail greyish and sparsely haired. **DISTRIBUTION** Subcoastal plains of north-western NT. **HABITAT AND HABITS** Lives in alluvial treeless plains along rivers, and adjacent mangroves and grassland. Forages at night for grasses and sedges. Shelters by day in soil cracks (dry season), or dense vegetation and shallow burrows (wet season).

Bush Rat ■ *Rattus fuscipes* TL 225–405mm, including tail 105–195mm
(Mootit)

DESCRIPTION Covered with soft, dense fur. Generally grey-brown and red-brown with light grey-brown underparts. Size varies significantly in different areas and between sexes; females are much smaller. Tail can be black, grey or brown and is slightly shorter than body. Ears rounded and pink-brown, whiskers long, and eyes dark and round. Feet can be white, brown, grey or pink; hindfeet often darker than forefeet. **DISTRIBUTION** Widely distributed and highly fragmented within coastal forests of NSW, Qld, Vic and inland forests within Vic, southern NSW and ACT. Also found in south-west corner of WA and a few isolated areas of SA, such as Victor Harbour to north of Adelaide, Kangaroo Island, and Port Lincoln. Can also occur on offshore islands close to east and south coasts. **HABITAT AND HABITS** Wide range of habitats including moist rainforests, wet and dry sclerophyll forests, woodland and heath to subalpine regions. Lives in dense forest understorey, sheltering in short burrows under logs or rocks and lining nests with grass. Prefers dense scrub and ground cover such as ferns, shrubs and rushes. Omnivore; eats fungi, grasses, lillies, fruits, seeds and insects.

RATS & MICE

Cape York Rat

■ *Rattus leucopus* TL 275–420mm, including tail 140–210mm (larger in north) (Rarrayn)

DESCRIPTION Greyish to blackish above, with hairs tipped with yellowish-brown, more yellowish on flanks, and obscure blackish patch around eyes. Tail brown, with varying amounts of white mottling, and scaled. Underparts pale grey to yellowish. Two subspecies show distinct differences in size and appearance. **DISTRIBUTION** Coastal north-eastern Qld. *R. l. leucopus* in northern Cape York Peninsula and *R. l. cooktownensis* from Cooktown to Paluma. **HABITAT AND HABITS** Found in rainforests. Omnivorous and nocturnal, foraging on the ground in leaf litter and other debris for insects, seeds, fruits, leaves and fungi. Shelters by day in burrows.

Swamp Rat

■ *Rattus lutreolus* TL 200–345mm, including tail 80–147mm (Koota)

DESCRIPTION Body fur dark grey to chestnut brown above and paler chestnut below. Feet have very dark pigments that produce an ashen colour. Ears are very small and sit within body fur line. Eyes are small and black, and don't protrude like Bush Rat (opposite). Tail is usually two thirds body length, dark brown to almost black and sparsely hairy. **DISTRIBUTION** Endemic to east and south-east Australia, ranging from Fraser Island in Qld, southwards through NSW and Vic to Kangaroo Island SA. Also present in the majority of Tas and Bass Strait Islands, and disjunct population in north-eastern Qld, between Atherton Tablelands and Paluma. **HABITAT AND HABITS** In mainland Australia, prefers wetland habitats such as swamps and along watercourses. Also prefers dense ground cover such as grassland, heath, dune scrub and sedges. Forms tunnels in the dense groundcover to move about. Cathemeral and forms nests in tunnels or burrows. In Tas, occurs in a wider range of habitats including alpine areas, wet sclerophyll forests, temperate moist forests and swampy moorlands of grass and sedges. Diet mainly grasses and sedges as well as insects.

▪ Rats & Mice ▪

Canefield Rat ▪ *Rattus sordidus* TL 225–370mm, including tail 100–160mm
(Minkala)

DESCRIPTION Fur coarse. Dark brown above, with numerous longer greyish-brown guard hairs, and pale grey below. Legs short and head has rounded 'Roman nose'. Tail dark grey, ringed and naked. **DISTRIBUTION** Broad coastal and inland band in eastern Qld, from Cape York Peninsula to Mackay Range. Possibly also on South-west Island NT, but no recent published records. **HABITAT AND HABITS** Lives in canefields, grassland, open grassy woodland and grassy areas in wetter forests. Nocturnal, sheltering by day communally in burrows, and foraging on the ground for plant material, mainly stems and leaves, supplemented with some insects.

Pale Field Rat ▪ *Rattus tunneyi* TL 200–345mm, including tail 80–150mm
(Djini; Chiiny Chiiny)

DESCRIPTION Medium-sized rodent with shiny but rough coat, pale yellow-brown and either grey or cream on underside. Large pale brown ears, large protruding eyes, whitish feet and tail shorter than head–body length with dark scale rings. **DISTRIBUTION** Mainly in Kimberley region, north-western WA, and northern NT, but also in western WA, southern, central and north-western NT. Also occurs on the east coast and inland from Cape York in Qld to Coffs Harbour NSW. Isolated areas in north-west NSW near Pilliga Scrub. **HABITAT AND HABITS** Found in various habitats in grassy open forests, cultivated pasture, cane-fields and pine plantations on sandy or rocky soils. Can also occur in wallum swamps and islands close to shore. Active and nocturnal. Forages on grass stems, seeds, roots, insects; rests during day in shallow burrows dug in loose, crumbly soil. Also builds runways in dense grass groundcover.

RATS & MICE/HARES & RABBITS

Long-haired Rat
■ *Rattus villosissimus* TL 220–400mm, including tail 100–180mm
(Mayaroo)

DESCRIPTION Greyish-brown above, washed with rufous, and with conspicuous longer black guard hairs along back, and greyish-white below. Tail slightly shorter than head-body, greyish and mostly naked. **DISTRIBUTION** Widely distributed from northern WA, through central and southern NT, western Qld and north-western SA, but range expands markedly in times of increased food availability. **HABITAT AND HABITS** Occurs mainly in riverine vegetation, but will move into other habitats after heavy rains. Normally active at night, but becomes more active in the day when food is in short supply, sheltering at other times in complex underground burrows. Plagues form after sustained periods of heavy rains.

LEPORIDAE, HARES AND RABBITS

European Brown Hare
■ *Lepus europaeus* TL 560–760mm, including tail 90–110mm

DESCRIPTION Fur above appears flecked, due to mix of tan, black and white hairs. Underparts white. Ears long, occasionally tipped with black; hind legs also elongated. Tail short and fluffy. The similar European Rabbit (see p. 110) is smaller with shorter ears and tail has brown upper surface. European Rabbits also run with tail up (showing white), while European Brown Hares run with tail down. **DISTRIBUTION** Introduced into Australia in Tas in 1830s, but failed to establish. Further introductions occurred on mainland and species became established in 1860s. Now widespread in continent's south-east, including Tas. **HABITAT AND HABITS** Occupies agricultural areas, scrubland, grassland and grassy woodland. Eats a range of vegetable matter including grasses, crops and bark. Largely solitary and predominantly nocturnal and crepuscular, sheltering during day in an open 'nest' in long grass.

HARES & RABBITS/PTEROPODIDS

European Rabbit
Oryctolagus cuniculus TL 390–515 mm, including tail 40–65mm

DESCRIPTION Fur typically grey-brown above and whitish grey below, but a few are light yellow-brown, red-brown, black or, rarely, white. Ears and hindlegs long; tail short

and fluffy. Underside of feet well furred; claws long and straight. Males may have a slightly broader head. Young rabbits have white star on forehead. The similar European Brown Hare (see p. 109) is larger with longer ears, longer hindlegs and running style; European Rabbits run with tail up (showing white), while hares run with tail down. **DISTRIBUTION** Wild populations introduced into Australia in mid to late 1800s and now widespread throughout all but monsoonal tropics. **HABITAT AND HABITS** Occurs in a variety of habitats, including open grassland, pasture and arid areas, but marked preference for habitats with low vegetation, deep well-drained soils and sufficient refuge sites, such as dense thickets or fallen

logs. Groups typically construct large, deep underground warrens (typically larger in more open country), used for breeding and shelter from higher temperatures, although may live above ground if sufficient dense groundcover exists. Herbivorous, eating a wide range of plant matter, but selectively grazes on the most nutritious parts. Generally gains sufficient moisture from food sources, but access to water is required in arid areas.

PTEROPODIDAE, OLD WORLD FRUIT-BATS

Bare-backed Fruit-bat
Dobsonia magna HB 190–230mm; FA 140–155mm

DESCRIPTION Large fruit-bat with blackish-brown fur on head and shoulders, and naked wings which, unlike in other fruit-bats, meet along midline of back rather than on sides, giving it a bare-backed appearance. Belly with sparse covering of paler greyish

brown fur. White claws on feet and thumb; second digit of wing clawless. Tail short. **DISTRIBUTION** Northern and eastern Cape York Peninsula, Qld and islands of Torres Strait. **HABITAT AND HABITS** Rainforests and tropical woodland, where it feeds at night on a variety of native and cultivated fruits, and some nectar. Shelters during day at the front of caves, and occasionally in dense vegetation and tree hollows. Flight slow and generally quiet, although wings touch together on top of upstroke, making a hollow 'pock' sound. Female has a single young, weaned after around 4 months. Often forages alone, and able to hover and manoeuvre into tight spaces within forest canopy.

◼ PTEROPODIDS ◼

Northern Blossom-bat
◼ *Macroglossus minimus* TL 49–67mm; FA 38–43mm

DESCRIPTION Pale reddish-brown fur, paler on belly, with males having pink V-shaped sternal gland on chest. Snout long and pointed. Tail short, with small flap of skin from base to ankles. **DISTRIBUTION** Tropical northern Australia, including offshore islands, from north-western WA, through northern NT and Cape York Peninsula Qld. **HABITAT AND HABITS** Lives in wet monsoon forests and woodland, bamboo thickets and paperbark swamps. Typically roosts alone or in small groups in dense foliage, emerging at night to feed predominantly on nectar and pollen, supplemented with some fruits. Although capable of hovering, normally lands on flowers to feed.

Eastern Tube-nosed Bat
◼ *Nyctimene robinsoni* TL 82–93mm; FA 65–69mm

DESCRIPTION Grey to reddish-brown fur, with narrow black dorsal stripe and yellow spots on wings and ears. Nostrils long and tubular. **DISTRIBUTION** Eastern Australia (including offshore islands), from far north-eastern NSW to northern Cape York Peninsula Qld. **HABITAT AND HABITS** Found in rainforests and wet sclerophyll forests. Feeds on variety of fruits and eucalypt blossoms, in forest canopy and understorey. Large wings enable it to hover and manoeuvre in tight spaces in the canopy. Mostly nocturnal, roosting by day within foliage in a regular roost or in the tree it was active in before dawn. Makes characteristic whistling call when flying.

◾ PTEROPODIDS ◾

Black Flying-fox
◾ *Pteropus alecto*
HB 185–280mm; FA 153–191mm

DESCRIPTION Predominantly black, occasionally with white flecks on belly fur, and often with reddish-brown fur on rear of neck. Eyes reddish, with faint reddish-brown eye-rings visible in some individuals. **DISTRIBUTION** Northern Australia (including offshore islands), from around Carnarvon WA, through northern NT, and northern and eastern Qld, to around Sydney NSW. Uncommonly recorded further south to Melbourne Vic. **HABITAT AND HABITS** Occurs in variety of wooded habitats, including rainforests, open forests, paperbark woodland, mangroves and bamboo thickets. Feeds at night on eucalypt blossoms, fleshy fruits, and occasionally leaves and tree sap.

Spectacled Flying-fox
◾ *Pteropus conspicillatus*
HB 225–247mm; FA 150–183mm

DESCRIPTION Fur predominantly black, with pale yellow to yellowish-brown on neck and shoulders, which extends to top of head and face in some individuals, and pale yellow rings around eyes. Exposed skin black. **DISTRIBUTION** Patchily distributed in north-eastern Qld (including offshore islands), from around Ingham in south to tip of Cape York Peninsula, and further north to New Guinea and nearby islands. **HABITAT AND HABITS** Roosts by day in large camps mainly within rainforests and riverine forests. Forages in broad range of habitats, including paperbark and eucalypt forests, mangroves and orchards, for variety of fruits, nectar, pollen and leaves.

PTEROPODIDS

Christmas Island Flying-fox
■ *Pteropus natalis* HB 150–200mm; FA 110–140 mm

DESCRIPTION Fur mainly dark brown, blacker on the head, and flecked with white (more so on head and belly). DISTRIBUTION Christmas Island, around 2,000km north-west Broome WA. HABITATS AND HABITS Found in rainforests, where it roosts during the day individually, in small groups or in large camps. Cathemeral. Forages in a range of habitats and feeds mainly on fruits, supplemented with some flowers and leaves. Over 35 species of plant have been recorded as food sources, and it may forage over 5km from its roost.

Grey-headed Flying-fox
■ *Pteropus poliocephalus*
HB 220–280mm; FA 152–177mm

DESCRIPTION Fur greyish-black on back, paler grey on face and belly, generally heavily flecked or dusted with silverish-grey, and with some reddish-brown on belly. Full collar of reddish-brown fur. DISTRIBUTION Coast, ranges and slopes of southern and eastern Australia, from around Mackay Qld, through NSW, ACT and Vic, to south-eastern SA. Scattered records further inland. HABITAT AND HABITS Roosts in large colonies of up to several thousand individuals in dense vegetation, favouring wetter areas and gullies. Nocturnal, foraging over wide area of native forests, gardens and orchards, for fruits, nectar, pollen and occasionally leaves.

PTEROPODIDS/GHOST BAT

Little Red Flying-fox
Pteropus scapulatus HB 120–200mm; FA 116–140mm

DESCRIPTION Fur pale to rich reddish-brown, often becoming more greyish on head, and occasionally with yellowish patch on neck and shoulders. Exposed skin reddish-brown, and wings largely transparent in flight. DISTRIBUTION Occurs in a broad coastal and inland band (including several islands) from central Vic, through ACT, NSW, Qld, NT and Kimberley region WA, becoming more coastal in western WA. HABITAT AND HABITS Lives in various habitats, from semi-arid woodland to rainforests and paperbark swamps. Feeds at night mainly on nectar and pollen, supplemented with fruits, sap and some insects. Roosts by day in large camps, often of several hundred thousand close to riparian zones such as rivers and creeks. Roosts in clusters very close together compared to other larger bats. Generally nomadic, in response to food availability, and readily roosts with other fruit-bat species.

MEGADERMATIDAE, GHOST BAT

Ghost Bat
Macroderma gigas
HB 98–118mm; FA 96–113mm

DESCRIPTION Pale brown or pale greyish fur above and greyish-white on underside, with skin of wing and tail membranes creamish-brown. Ears large, joining together in middle, and nose with fleshy leaf shape on top of snout. Generally darker closer to coast. DISTRIBUTION Widespread but heavily fragmented, with subpopulations throughout northern Australia. HABITAT AND HABITS Occurs in range of habitats, including rainforests and grassy woodland. Roosts by day in colonies in caves and rocky crevices, emerging at night to feed on small vertebrates, such as birds, mammals (including Orange Leaf-nosed Bat, see p. 117), reptiles, frogs and large insects.

HORSESHOE BATS/LEAF-NOSED BATS

RHINOLOPHIDAE, HORSESHOE BATS

Eastern Horseshoe-bat
Rhinolophus megaphyllus HB 44–53mm; FA 45–52mm

DESCRIPTION Fine, moderately long fur, usually greyish-brown above (reddish-orange in some northern individuals), with white tips giving somewhat frosted appearance, and slightly paler on belly. Naked skin of wings, ears and nose grey tinged with pinkish. DISTRIBUTION Coast and eastern slopes of eastern mainland Australia, from northern Qld, through eastern NSW to central Vic. HABITAT AND HABITS Inhabits rainforests, sclerophyll forests, open woodland and grassland. Roosts by day in caves, disused mines, tunnels, tree hollows and buildings, and emerges to forage nocturnally on insects, mainly moths.

John Harris/Wildlife Experiences

HIPPOSIDERIDAE, LEAF-NOSED BATS

Dusky Leaf-nosed Bat
Hipposideros ater HB 33–46mm; FA 34–41mm

DESCRIPTION Tiny. Fur long and soft, normally light greyish-brown, but can be more orange. Skin of wings and tail membrane blackish, with tail fully enclosed in tail membrane. DISTRIBUTION Northern Australia, from Kimberley region WA, through northern NT and Cape York Peninsula, to around Townsville Qld. HABITAT AND HABITS Inhabits rainforests, open forests, woodland, mangroves and grassland, but favours dense vegetation. Roosts by day either singly or in small groups in caves, abandoned mines and tree hollows, and forages almost exclusively for moths, which are caught in flight below canopy height. Prefers to roost in darker and more humid parts of roost site.

Ryan Francis

LEAF-NOSED BATS

Diadem Leaf-nosed Bat
■ *Hipposideros diadema*
HB 74–96mm; FA 77–85mm

DESCRIPTION Large leaf-nosed bat. Fur generally greyish or pale brown above to dark reddish-brown, with conspicuous white patch on shoulders and back. DISTRIBUTION Coast and ranges of north-eastern Qld, from Townsville in south, north to around Iron Range on eastern Cape York Peninsula. Australian subspecies, *H. d. reginae*, one of 15 widespread subspecies. **HABITAT AND HABITS** Inhabits lowland rainforests, forests and woodland. Roosts by day either solitarily or in small to large groups in large caves and abandoned mines, and occasionally in dense vegetation or tree hollows, emerging at night to forage mostly for invertebrates and some small vertebrates.

Northern Leaf-nosed Bat
■ *Hipposideros stenotis*
HB 39–46mm; FA 42–46mm

DESCRIPTION Fur thin and long, pale greyish-brown above and slightly paler below, with white on base of each strand. DISTRIBUTION Northern Australia, from Kimberley region, including numerous offshore islands, WA, through northern NT to around Mt Isa in north-western Qld. **HABITAT AND HABITS** Occurs in rocky sandstone hills and escarpments in close proximity to open forests, woodland and grassland. Roosts by day individually or in small groups in small caves, disused mines and rocky crevices. Insectivorous, foraging close to the ground and above creek lines at night for flying insects, including moths and beetles.

ORANGE LEAF-NOSED BAT/SHEATH-TAILED BATS

RHINONICTERIS, ORANGE LEAF-NOSED BAT

Orange Leaf-nosed Bat
■ *Rhinonicteris aurantia*
HB 45–55mm; FA 45–50mm

DESCRIPTION Fur generally uniform rich orange, but can also be whitish or yellowish-brown. Wing and tail membranes dark brown. **DISTRIBUTION** Northern and western Australia from Pilbara and Kimberley region, including several offshore islands, WA, through northern NT to far north-western Qld. **HABITAT AND HABITS** Lives in rocky areas. Roosts by day in warm (28–32 °C), humid (more than 95 per cent) dark caves and disused mines in colonies of up to 20,000 individuals, although numbers are normally much lower. Emerges at dusk to feed close to the ground on flying insects, including moths (mainly), beetles and termites, which are caught using skillful manoeuvring during fast flight.

EMBALLONURIDAE, SHEATH-TAILED BATS

Yellow-bellied Sheath-tailed Bat
■ *Saccolaimus flaviventris* HB 72–92mm; FA 66–82mm

DESCRIPTION Glossy black fur above and white to creamish-yellow below. Head flattened, with naked, dark pinkish-brown snout and moderately long, triangular ears. **DISTRIBUTION** Throughout northern and eastern Australia, including northern WA, NT, northern and eastern SA (largely migratory), Qld, NSW and Vic. Also offshore islands with suitable habitat. **HABITAT AND HABITS** Occupies wide variety of habitats within broad range, including forests, woodland, grassland and deserts. Roosts in small to medium-sized groups in large tree hollows, and forages above canopy height or in open spaces at fringes for insects.

◾ SHEATH-TAILED BATS ◾

Bare-rumped Sheath-tailed Bat
◾ *Saccolaimus saccolaimus*
HB 81–99mm; FA 72–80mm

DESCRIPTION Fur reddish-brown above with scattered white patches, and pale grey below, with fur of back not extending past hips. Wing membranes olive-brown. **DISTRIBUTION** Mainly near coastal areas and islands of northern Australia, from Kimberley Region WA, through northern NT, to just south of Townsville Qld. **HABITAT AND HABITS** Favours eucalypt forests and woodland. Roosts by day in small to medium-sized groups (recorded up to 77) in tree hollows, and feeds at night on flying insects, mainly by foraging just above the canopy, but may also hunt close to the ground.

Coastal Sheath-tailed Bat
◾ *Taphozous australis* HB 61–75mm; FA 61–68mm

DESCRIPTION Fur pale brown to greyish-brown above and paler grey below. Wing membranes reddish-brown, with pouches. Male has throat-pouch, which is much reduced in female to a ridge of bare skin. **DISTRIBUTION** Coast and islands of eastern Qld, from northern Cape York Peninsula, south to Shoalwater Bay. Also recorded on islands in Torres Strait and New Guinea. **HABITAT AND HABITS** Roosts by day in small colonies in caves and rocky crevices, mostly along coastal cliffs. Emerges at night to forage above the canopy of rainforests, forests, mangroves, heathland and swamps for beetles and other flying insects.

Sheath-tailed Bats

Common Sheath-tailed Bat
■ *Taphozous georgianus* HB 62–80mm; FA 62–73mm

DESCRIPTION Fur dark brown above, paler brown below, and becoming more creamish towards tail. Wings semi-transparent in flight. **DISTRIBUTION** Northern Australia, from Pilbara and Kimberley regions WA, through northern NT to north-western Qld. **HABITAT AND HABITS** Inhabits wet and dry forests, woodland and spinifex grassland. Feeds in open spaces, generally along ridge tops and rocky outcrops, at night on insects (mainly beetles). Roosts by day in small to moderately sized colonies in caves, generally close to cave entrances on more vertical walls.

Hill's Sheath-tailed Bat ■ *Taphozous hilli* HB 65–81mm; FA 60–72mm

DESCRIPTION Fur dark brown above, washed with grey, and only marginally paler below, but fur on belly has paler base. Wing and tail membranes brown. **DISTRIBUTION** Arid inland Australia in broad band through central WA, central and southern NT, and northern SA. **HABITAT AND HABITS** Occurs in drier open woodland, shrubland, grassland and open plains, in areas with suitable rocky places and adequate caves and crevices. Roosts in caves, abandoned mines and rocky crevices, and forages at night, presumably for insects.

SHEATH-TAILED BATS/FREE-TAILED BATS

Troughton's Sheath-tailed Bat
■ *Taphozous troughtoni*
HB 79–93mm; FA 67–76mm

DESCRIPTION Fur olive-brown, and paler brownish on face and inside ears. Tail fully enclosed in membrane. Difficult to distinguish from very similar, and wider ranging, Common Sheath-tailed Bat (see p. 119). The two species are known to roost together and were considered conspecific until 2008. **DISTRIBUTION** Central and eastern Qld, from southern Cape York Peninsula in north, south to Taroom and west to around Mt Isa. **HABITAT AND HABITS** Occurs in rocky areas in open woodland and grassland. Roosts by day in caves, disused mines and rocky crevices, always within twilight zone, close to entrance. Emerges at dusk to forage above canopy for flying insects, including grasshoppers.

MOLOSSIDAE, FREE-TAILED BATS

White-striped Free-tailed Bat
■ *Austronomus australis* HB 85–100mm; FA 57–65mm

DESCRIPTION Dark reddish-brown to blackish above and only slightly paler below, with distinct white patches at junction with wing and occasionally on chest. **DISTRIBUTION** Throughout Australia, except tropical north of WA, NT and Qld. Vagrant in Tas. **HABITAT AND HABITS** Inhabits forests, woodland, grassland and deserts with scattered trees. Roosts by day, either solitarily or in small colonies, in tree hollows, emerging at night to forage for moths, beetles and other insects, normally caught in flight, but also on the ground. Echolocation frequency quite audible and of lower range than any other Australian bat call (11–13 kHz).

FREE-TAILED BATS

Northern Free-tailed Bat
Ozimops lumsdenae HB 54–68mm; FA 37–41mm

DESCRIPTION Fur brown to orange-brown above, tinged with yellow on sides of neck, and only slightly paler on undersurface. Skin on wings, ears and muzzle blackish-brown. Similar to Greater Northern Free-tailed Bat, but forearm length smaller. **DISTRIBUTION** Coastal and inland areas of northern Australia, from Pilbara region WA, through northern NT, to southern Qld. **HABITAT AND HABITS** Found in variety of habitats, including rainforests, shrubland and grassland. Roosts by day in tree hollows and buildings, emerging at night to forage.

Southern Free-tailed Bat
Ozimops planiceps HB 54–55mm; FA 32–36mm

DESCRIPTION Fur pale brown to greyish-brown on head and back, and paler brown on underside, with white bases to individual hairs. Skull flattened. **DISTRIBUTION** Western slopes of Great Dividing Range, from northern NSW, through central Vic, to Flinders and Gawler Ranges of eastern SA. **HABITAT AND HABITS** Occupies forests and woodland, including box-ironbark, mallee, Mulga and River Red Gum, with annual rainfall of 300–700mm. Roosts by day in tree hollows and in buildings, emerging to forage for insects.

◾ Free-tailed Bats ◾

Ride's Free-tailed Bat
◾ *Ozimops ridei*
HB 42–55mm; FA 30–35mm

DESCRIPTION Variable. Fur pale brownish-grey to dark brown above, and paler below. Head broad, with large ears that almost meet in middle of forehead. Difficult to distinguish from **Inland Free-tailed Bat** *O. pertersi* and Southern Free-tailed Bat (see p. 121) in areas where ranges overlap. DISTRIBUTION Riverine systems and drainages of eastern Australia and islands of Torres Strait, from far northern Qld, through eastern NSW and along Murray River Vic, and south-eastern SA. HABITAT AND HABITS Lives in wetter forests, open woodland, riverine vegetation and swampland. Roosts by day in tree hollows and tree crevices, under loose bark and in buildings, occasionally with other bat species, emerging to forage at night for insects.

Bristle-faced Free-tailed Bat
◾ *Setirostris eleryi*
HB 41–50mm; FA 31.5–36mm

DESCRIPTION Fur yellowish-brown to greyish-brown above and paler below. Head has long, narrow, sparsely haired snout, with series of 26–30 conspicuous dark bristles on snout, and triangular-shaped ears. DISTRIBUTION Scattered records from southern NT, northern SA, northern and western NSW, and southern and central Qld. HABITAT AND HABITS Poorly known. Recorded foraging at night, quite low to the ground near watercourses in open woodland. Shelters by day, either individually or in small groups, in tree hollows and fissures (cracks, splits and depressions).

■ Bent-winged Bats ■

Miniopteridae, Bent-winged Bats

Little Bent-winged Bat
■ *Miniopterus australis* HB 40–45mm; FA 37–41mm

DESCRIPTION Dark brown above and paler below. Tip of wing has a long joint of the third finger, which folds back and is bent under the wing while bat is sleeping. Fur is quite long and thick on crown and neck. Muzzle short and ears triangular. Distinguished from **Common Bent-winged Bat** M. *schreibersii* by smaller size. **DISTRIBUTION** From Cape York in Qld, down to Wollongong in NSW. **HABITAT AND HABITS** Occurs within rainforest, sclerophyll forests, paperbark swamps, banksia scrub and dense coastal forests. Forages for insects below forest canopy and within very densely vegetated habitats. Readily roosts during day with other species in tunnels, caves, culverts, under bridges, tree hollows, stormwater drains, mines and buildings, often in quite large numbers.

Large Bent-winged Bat
■ *Miniopterus orianae* HB 47–65mm; FA 42–50mm

DESCRIPTION Currently recognized subspecies (M. *o. bassanii*, M. *o. oceanensis* and M. *o. orianae*) may form three geographically distinct species. Fur uniformly greyish-brown to dark brown (*bassanii*), reddish-brown to dark brown (*oceanensis*) or dark brown (*orianae*) on back, paler below. Tip (phalanx) of third finger 3–4 times longer than second phalanx, and bent under wing when at rest. Head dome shaped, with short snout and short, broad and triangular ears with rounded tips. Wing and tail membranes dark brown to blackish. **DISTRIBUTION** Northern and eastern Australia, from southern Kimberley region WA, through northern NT (*orianae*); Cape York Peninsula, northern Qld, through eastern NSW, to eastern Vic (*oceanensis*); and south-western Vic to south-eastern SA (*bassanii*). **HABITAT AND HABITS** Lives in variety of wet and dry wooded habitats. Roosts colonially by day in caves, emerging at dusk to forage for flying insects, especially moths.

VESPERTILIONIDS

VESPERTILIONIDAE, VESPERTILIONID BATS

Golden-tipped Bat ■ *Phoniscus papuensis* HB 50–60mm; FA 35–40mm

DESCRIPTION Fur long and curly, dark brown to blackish, and tipped with golden-yellow. Golden fur extends onto forearms, legs, long tail and tail membrane. Head has long, pointed nose, tall, rounded crown and funnel-shaped ears. **DISTRIBUTION** Coastal and near coastal eastern Australia, from northern Cape York Peninsula to far south-eastern NSW. **HABITAT AND HABITS** Found in rainforests, wet and dry sclerophyll forests, and mangroves, generally with a dense understorey. Roosts by day in disused dome-shaped birds' nests, but also in dense vegetation or tree hollows, and forages at night for spiders (almost exclusively) and some insects.

Arnhem Long-eared Bat

■ *Nyctophilus arnhemensis* HB 40–50mm; FA 33–40mm

DESCRIPTION Fur brown to reddish-brown above, darker on bases of hairs, and only marginally paler below. Head blunt. Shares range with similar Pygmy Long-eared Bat (see p. 128), which has shorter ears and more contrast between colour of back and belly, and larger Pallid Long-eared Bat (see p. 126). **DISTRIBUTION** Tropical northern Australia, including offshore islands, from Exmouth WA, through northern NT, to far north-western Qld. **HABITAT AND HABITS** Lives in rainforests, forests, open grassy woodland and mangroves. Roosts by day under loose bark or in dense foliage, emerging at dusk to forage slowly among dense vegetation mainly for insects, including beetles and termites, and some spiders.

VESPERTILIONIDS

Eastern Long-eared Bat ◼ *Nyctophilus bifax* HB 35–55mm; FA 38–47mm

DESCRIPTION Fur orange-brown above, normally paler below. Head blunt, with shallow, centrally depressed ridge across snout. Ears long. **DISTRIBUTION** Eastern mainland and adjacent islands, from northern Cape York Peninsula Qld to north-eastern NSW.

HABITAT AND HABITS Lives in rainforests and riverine forests, and less commonly in open forests and woodland. Roosts either singly or in small groups in tree hollows, under loose bark, in dense foliage and occasionally in buildings. Forages at night, mainly for moths, by perching in foliage and waiting to detect prey close by before seizing it in short flights.

Corben's Long-eared Bat
◼ *Nyctophilus corbeni* HB 55–65mm; FA 38–45mm

DESCRIPTION Uniform dark grey washed with pale brown, and with greyish wing and tail membranes. Head has blunt snout (with low ridge) and large, broad ears. **DISTRIBUTION** South-eastern mainland, including southern inland Qld, central and south-western NSW, northern Vic and south-eastern SA. **HABITAT AND HABITS** Occurs in large woodland with dense understorey, and similar wooded areas. Roosts mostly solitarily by day in tree hollows and under loose bark, emerging to forage in the understorey and adjacent open spaces, for insects. Often flies close to the ground when foraging.

VESPERTILIONIDS

Pallid Long-eared Bat ▪ *Nyctophilus daedalus* HB 53–57mm; FA 38–46mm

DESCRIPTION Fur pale yellowish-brown above, paler brown below. Head blunt, with shallow, centrally depressed ridge across snout. **DISTRIBUTION** Two distinct ranges. Pilbara region WA, and northern Australia, from Kimberley region WA, through northern NT (and offshore islands), to far north-western Qld.

HABITAT AND HABITS Mainly found in wetter forests, but also in tall, open forests and open woodland. Roosts by day in tree hollows, under loose bark or in dense foliage. Forages at night, mainly for beetles, by perching in foliage and waiting to detect prey close by before seizing it in short flights or gleaning it from foliage.

Lesser Long-eared Bat ▪ *Nyctophilus geoffroyi* HB 38–50mm; FA 32–42mm

DESCRIPTION Fur long and soft, pale greyish-brown above and paler brown or whitish below, with individual hairs with dark base. **DISTRIBUTION** Widespread throughout mainland, Tas and offshore islands, but seemingly absent from coastal north-eastern Qld.

HABITAT AND HABITS Found in almost all habitats, from rainforests and mangroves, to alpine woodland and arid deserts. Roosts either singly or in small groups of 2–3 in tree hollows, under bark or in crevices in trees or rock faces, and similar sites. Emerges at night to forage above the understorey and often close to the ground for insects (mainly moths and crickets) and other invertebrates.

VESPERTILIONIDS

Gould's Long-eared Bat ■ *Nyctophilus gouldi* HB 44–52mm; FA 36–48mm

DESCRIPTION Fur greyish-brown to dark grey above, and paler grey with light brown wash below. Head blunt, with mid-sized ridge across snout, and very long ears that fold down when at rest, with a small triangular lower lobe that extends over ear opening. Wings short and wide. **DISTRIBUTION** Occurs within eastern and south-western Australia, from north-eastern Qld, through coastal and inland NSW, ACT and Vic, including along Murray River, to far south-eastern SA. Also south-western WA. **HABITAT AND HABITS** Inhabits rainforests, forests and woodland, preferring wetter areas, where it roosts during day, either solitarily (mainly males) or in small colonies in tree hollows, tree crevices and under loose bark. Emerges at dusk to forage for insects, mainly moths, and other invertebrates, which it catches either by sight or echolocation. Female is able to breed at around 8 months of age, and gives birth to 1 or 2 young. Goes into torpor in cooler months.

Tasmanian Long-eared Bat
■ *Nyctophilus sherrini* HB 55mm; FA 44–48mm

DESCRIPTION Fur dark brown above and paler brown below. Head large, with shallow, centrally depressed ridge across broad snout. Larger in size than Lesser Long-eared Bat (see opposite), which has distinctive Y-shaped groove on snout. Has the broadest wings of any bat species in Tas. **DISTRIBUTION** Tas, from sea level to about 740m. **HABITAT AND HABITS** Occurs in mature forests, woodland, swamps and shrubland. Roosts by day in tree hollows and crevices, and under loose bark, emerging at night to forage in dense vegetation, close to foliage or on the ground, mainly for caterpillars and scorpions, but also feeds on moths.

◾ Vespertilionids ◾

Pygmy Long-eared Bat ◾ *Nyctophilus walkeri* HB 38–44mm; FA 30–36mm

DESCRIPTION Fur pale brown above, normally tinged with orange, and more cream to pale brown below, with contrasting blackish wing membranes. Shares range with similar but larger Arnhem Long-eared Bat (see p. 124), which is more uniform in colour.

DISTRIBUTION Northern Australia, from northern Kimberley region WA, through northern NT to far north-western Qld. **HABITAT AND HABITS** Occurs in rainforests, forests, grassy woodland and shrubland, usually in vicinity of water. Roosts in large colonies in dense vegetation or on palm leaves, and emerges at dusk to forage for insects, mainly moths and beetles, and other invertebrates.

Gould's Wattled Bat ◾ *Chalinolobus gouldii* HB 46–60mm; FA 36–47mm

DESCRIPTION Predominantly dark brown fur on back, with almost black shoulders, neck and head. 'Wattled' refers to fleshy lobe or wattle that sits between base of ear and corner of mouth. **DISTRIBUTION** Found in nearly every climatic zone within Australia. **HABITAT AND HABITS** One of Australia's most commonly encountered bats, occurs in most habitats with trees. Roosts in hollows of old trees and in some cases in old buildings. In southern parts of Australia in cooler climate, usually one of the first bats to emerge from

its winter hibernation, and one of the first bats to emerge at dusk each evening. A capable flier, much of its hunting is done above canopy but individuals will use 'flyways' and paths within forest to ease their movement through the landscape. Diet includes moths, cockroaches, stoneflies, crickets, cicadas and many other flying and non-flying insects.

VESPERTILIONIDS

Chocolate Wattled Bat ■ *Chalinolobus morio* HB 40–52mm; FA 33–42mm

DESCRIPTION Fur soft and thick, chocolate-brown above and below, and sometimes paler on belly in inland population. Top of head rounded, and ears broad and relatively short. **DISTRIBUTION** Broad, but patchy through coastal and inland Australia, including Tas and several offshore islands, from around Townsville Qld in north-east, to Pilbara region WA in north-west. **HABITAT AND HABITS** Found in large rainforests, sclerophyll forests, woodland and shrubland, mostly along watercourses in drier inland. Roosts mainly in trees, including in hollows and under loose bark, but also in caves and inside buildings. Forages mostly between the understorey and canopy.

Hoary Wattled Bat ■ *Chalinolobus nigrogriseus* HB 40–48mm; FA 32–37mm

DESCRIPTION Fur greyish-black on back and greyish-brown on belly, with most individuals having white tips on each hair, giving frosted (hoary) appearance, which is more pronounced in west of range. **DISTRIBUTION** Northern and eastern Australia (including offshore islands), from southern Kimberley region WA, through northern NT, northern and eastern Qld, to around Coffs Harbour NSW. **HABITAT AND HABITS** Found in rainforests, wet and dry forests, open woodland, grassland and sand dunes. Roosts by day in tree hollows, and less commonly in rock crevices, emerging at dusk to forage along watercourses and above the tree canopy for beetles, moths and other invertebrates.

▪ Vespertilionids ▪

Little Pied Wattled Bat
▪ *Chalinolobus picatus* HB 41–49mm; FA 31–38mm

DESCRIPTION Fur long and glossy black above, extending onto tail membrane, where it becomes more brownish. Blackish below, with greyish wash, and fur alongside wing and tail membranes white, forming broad white V-shape. **DISTRIBUTION** Eastern inland, from central eastern Qld, through central and western NSW, to north-western Vic and eastern SA. **HABITAT AND HABITS** Occurs in open forests, open woodland and shrubland, and often in association with waterways. Roosts individually or in small groups of up to 50–100 individuals, but more normally up to 50 individuals, mainly in caves, disused mines and tree hollows, and under tree bark. Emerges at dusk to forage for insects (mostly moths).

Eastern Falsistrelle
▪ *Falsistrellus tasmaniensis* HB 57–66mm; FA 45–56mm

DESCRIPTION Relatively large with long, fine, dark brown to reddish-brown fur (paler and more greyish below). Ears long, slender and narrow, set well back on head. Nose has some scattered hair. Distinguished from Greater Broad-nosed Bat (opposite) by having two pairs of upper incisors. **DISTRIBUTION** Occurs in south-eastern Australia, from far south-eastern SA, through Vic, Tas (including islands of Bass Strait), and eastern NSW to south-eastern Qld. **HABITAT AND HABITS** Inhabits rainforests, wet sclerophyll and tall open eucalypt forests, generally with a dense understorey. Roosts mainly in hollow eucalyptus tree trunks, under loose bark on trees, but has also been recorded in caves, in small to medium-sized colonies of up to 80 individuals. Forages nocturnally above forest canopy, and creeks and tracks just below canopy height for large moths, beetles, weevils and other insects.

◾ Vespertilionids ◾

Greater Broad-nosed Bat
◾ *Scoteanax rueppellii* HB 63–73mm; FA 51–56mm

DESCRIPTION Fur long and fine, dark brown to reddish-brown above and marginally paler below. Head broad and short square muzzle, with shallow forehead. Larger than other broad-nosed bats. Similar to Eastern Falsistrelle (opposite) but has two upper incisors not four. **DISTRIBUTION** Throughout coast and ranges of eastern Australia, from Atherton Tablelands, north-eastern Qld, south to north-eastern Vic. Northern Qld subpopulation separated from main south-eastern population. In NSW occurs abundantly in New England Tablelands below 500m. Although widely distributed, and thought to have a large overall population size, considered to be thinly scattered throughout its range. **HABITAT AND HABITS** Found in rainforests, wet sclerophyll forests, dry open forests and woodland, preferring wetter gullies and waterways on the Great Dividing Range and to the coast. Roosts by day in tree hollows, crevices and under loose bark, and forages at night for insects and other bat species usually along creeks and river corridors.

Inland Broad-nosed Bat
◾ *Scotorepens balstoni* HB 42–60mm; FA 32–41mm

DESCRIPTION Fur pale greyish-brown to yellowish-orange, and paler brown below; fur has paler base and darker tips. Snout square shaped and sparsely furred, with V-shaped cleft in lower lip. **DISTRIBUTION** Broadly distributed throughout arid and semi-arid inland Australia, reaching coast only in western WA and south-eastern SA. **HABITAT AND HABITS** Lives in woodland, shrubland and grassland. Forages over waterways between dusk and dawn for insects, including mosquitoes, beetles, cockroaches, termites and ants. Roosts at other times in colonies of up to 45 mainly within tree hollows.

VESPERTILIONIDS

Little Broad-nosed Bat ■ *Scotorepens greyii* HB 37–53mm; FA 27–35mm

DESCRIPTION Fur long and fine, greyish-brown to reddish-brown above, paler below. Head broad, with shallow forehead, and snout has slightly raised nostrils. Reddish individuals closely resemble Northern Broad-nosed Bat (opposite). **DISTRIBUTION** Most

of Australia, but absent from Cape York Peninsula Qld, south-east mainland and Tas, southern and western SA, and most of southern WA. **HABITAT AND HABITS** Occurs in various wooded habitats, including closed forests, open forests, open woodland and shrubland. Typically forages close to the tree canopy near water for beetles and other insects, sometimes almost as large as itself. Roosts by day in tree hollows or similar spaces, in small colonies of up to about 20 individuals.

Eastern Broad-nosed Bat
■ *Scotorepens orion* HB 44–53 mm; FA 32–39 mm

DESCRIPTION Fur rich brown above, paler below, and bare skin blackish. Appears darker

and more heavily-built than other broad-nosed bats. Ears elongated with rounded tips and tail fully enclosed in membrane. **DISTRIBUTION** Occurs along Great Dividing Range and adjacent coastal plains within south-eastern mainland, from south-eastern Qld to southern Vic. **HABITAT AND HABITS** Typically inhabits tall wet sclerophyll forests and rainforests, where it roosts during the day in tree hollows and, uncommonly, in buildings. Also recorded in open forests near ACT.

VESPERTILIONIDS

Northern Broad-nosed Bat
■ *Scotorepens sanborni* HB 37–52mm; FA 28–36mm

DESCRIPTION Fur long and fine, brown above, washed with reddish, and paler below. Head broad, with shallow forehead and wide, rounded ears that have thin fleshy lobe passing partially over opening. Snout dark and mostly naked, with slightly raised nostrils. Wing and tail membranes blackish. Closely resembles reddish individuals of Little Broad-nosed Bat (opposite). **DISTRIBUTION** Two distinct regions of northern Australia, in north-eastern Qld, and from north-western WA to north-western NT. **HABITAT AND HABITS** Occurs in open woodland and heathland in east, and paperbark forests and mangroves in west. Normally roosts in small colonies of up to 20 individuals in tree hollows, but several hundred may gather together to roost in roof spaces of buildings. Forages for beetles and other insects either in understorey (east of range), or in canopy area (west).

Inland Forest-bat ■ *Vespadelus baverstocki* HB 35–44 mm; FA 27–31 mm

DESCRIPTION Fur generally yellowish-brown above, with grey bases to hairs, and creamish to fawn below, but can be uniformly greyish-brown in some individuals. **DISTRIBUTION** Broadly distributed in inland Australia. Found in arid and semi-arid parts of all mainland states. **HABITAT AND HABITS** Lives in woodland, shrubland and grassland. Roosts by day in tree hollows and buildings, often in colonies of about 50–100 individuals. Feeds at night, presumably mainly on moths and other insects.

VESPERTILIONIDS

Northern Cave-bat
- *Vespadelus caurinus*
HB 32–40mm; FA 27–32mm

DESCRIPTION Fur greyish-brown, browner on rump, and with darker blackish bases to hairs. Wing and tail membranes blackish-brown. May be confused with the larger **Yellow-lipped Cave-bat** *V. douglasorum*, which is also found in the Kimberley region of WA. **DISTRIBUTION** Northern Australia, from Kimberley region WA, through northern NT (including offshore islands), to western Gulf of Carpentaria Qld. **HABITATS AND HABITS** Found in rocky areas, often close to, or within, forests, woodland and grassland, where it roosts in small groups within rocky crevices. Feeds during dusk and at night on flying insects, caught within the tree canopy.

Large Forest-bat ▪ *Vespadelus darlingtoni* HB 38–49mm; FA 33–37mm

DESCRIPTION Fur long and reddish-brown, becoming darker brown to blackish with age; pale brown tips to hairs. Head somewhat blunt, with short, rounded dark ears.
DISTRIBUTION Ranges and slopes of south-eastern Australia, and offshore islands

(including Lord Howe Island), from south-eastern Qld, through eastern NSW and most of Vic (absent from north-west), to south-eastern SA and Tas. **HABITAT AND HABITS** Lives in rainforests, forests, woodland and swampland. Roosts by day in colonies of up to 80 individuals, and forages in dense vegetation and between understorey and canopy for range of invertebrates, including moths, beetles and ants.

■ VESPERTILIONIDS ■

Finlayson's Cave-bat
■ *Vespadelus finlaysoni* HB 34–46mm; FA 30–38mm

DESCRIPTION Fur long (extending onto snout), brown to blackish above, washed with reddish or yellowish, and slightly paler brown below. Skin of face, ears and wing and tail membranes blackish. **DISTRIBUTION** Arid and semi-arid Australia, including central and western WA, northern SA, western Qld and most of NT (including Top End and offshore islands). **HABITAT AND HABITS** Occupies grassland, grassy woodland and open forests. Roosts by day in small to moderately large colonies in fronts of caves, abandoned mines and rocky crevices, and emerges at dusk to forage in vegetation and near waterholes.

Eastern Forest-bat ■ *Vespadelus pumilus* HB 35–44mm; FA 28–33mm

DESCRIPTION Fur long and thick, and dark brown above with black bases to hairs, and paler below, extending onto tail membrane. **DISTRIBUTION** Patchily distributed along coast and ranges of eastern Australia, from Atherton Tablelands Qld to around Sydney NSW. **HABITAT AND HABITS** Lives in wetter mature forests (including rainforests and sclerophyll forests) and bunya pine plantations. Roosts by day in tree hollows, with males generally roosting alone and females in larger groups. Forages between understorey and tree canopy for moths and other insects. Observed entering state of torpor on daily basis in morning and afternoon, even during optimal weather conditions and food availabilty.

▪ Vespertilionids ▪

Southern Forest-bat ▪ *Vespadelus regulus* HB 36–39 mm; FA 28–34 mm

DESCRIPTION Fur brown to grey above, washed with reddish, and greyish below, with dark brown bases to hairs. Ears and wing membranes grey. **DISTRIBUTION** Southern Australia, from south-eastern Qld, through eastern NSW, Vic, Tas and southern SA, to south-western WA. **HABITAT AND HABITS** Found in rainforests, wet and dry sclerophyll forests, and open woodland. Roosts by day in tree hollows, generally in colonies of up to 100 individuals. Feeds at night on moths and other insects, which are caught close to the ground. Common visitor to houses adjacent to suitable habitats, in parts of its range.

Eastern Cave-bat
▪ *Vespadelus troughtoni*
HB 38–44mm; FA 32–36mm

DESCRIPTION Fur pale brown above, tipped with reddish-brown (mostly on head and face), and paler below, with darker bases and paler tips to individual hairs. **DISTRIBUTION** Broad coastal and inland band in eastern Australia, from around Cooktown Qld to central-eastern NSW. Isolated records from further inland, suggesting wider distribution. **HABITAT AND HABITS** Found in forests and woodland in close proximity to sandstone or volcanic escarpments. Roosts by day in rocky areas, including shallow caves, disused mines and piles of boulders. Forages for small insects, including mosquitoes, over waterways, and within a small area.

◾ VESPERTILIONIDS ◾

Little Forest-bat ◾ *Vespadelus vulturnus* HB 35–48mm; FA 24–33mm

DESCRIPTION Fur pale grey to brownish above, and paler grey below, occasionally streaked with white. Head somewhat blunted, with raised forehead. **DISTRIBUTION** South-eastern Australia, from central southern Qld, through most of NSW (absent from north-west), ACT, and most of Vic to south-eastern SA and Tas (including Flinders Island). **HABITAT AND HABITS** Occurs in variety of wooded habitats, often near water. Roosts by day in colonies of up to 120 individuals in tree hollows with small entrances. Forages at night just below canopy and in open spaces between trees, for moths, beetles and other flying arthropods.

Large-footed Myotis
◾ *Myotis macropus* HB 35–50mm; FA 37–43mm

DESCRIPTION Fur dark greyish-brown to reddish-brown above and yellowish-brown below, with hairs on chest tipped with grey. Feet large with hook-shaped claws. **DISTRIBUTION** Coastal areas, inland rivers, creeks and lakes in northern and eastern Australia, from Kimberley region WA, through northern NT, northern and eastern Qld, eastern and southern NSW, ACT, and central Vic, to south-eastern SA. **HABITAT AND HABITS** Occurs in various habitats near water. Roosts by day in caves, tree hollows, culverts, under bridges and in dense vegetation, and in human structures near water, such as tunnels and bridges. Feeds at night on aquatic insects and small fish. Can hibernate through winter months in colder parts of range.

Dogs

CANIDAE, DOGS

Dingo
■ *Canis familiaris* TL 1.1–1.35m, including tail 260–360mm

DESCRIPTION Top of body sandy-yellow to red-brown, occasionally dark brown to black. Underside lighter, from white to tan. Hybrids with domestic dogs are common but cannot be accurately distinguished visually. **DISTRIBUTION** Introduced. Historically, found across mainland Australia. Now occurs across northern Australia, north-west SA, and down east coast to Gippsland region Vic. Distribution somewhat restricted due to man-made Dingo fence constructed to keep Dingoes out of much of agricultural land in SA, southern Qld and NSW. **HABITAT AND HABITS** Found in wide variety of habitats, from cool mountainous forests to deserts. Active day and night. Opportunistic hunter, normally hunting alone, but will also form small packs, working together to take down large prey. A variety of prey species is taken, including sheep and young cattle, which has led to them being shot and poisoned by landholders. Will also feed on carrion.

Red Fox
■ *Vulpes vulpes* TL 960–1190mm, including tail 360–450mm

DESCRIPTION General coat colour can vary, with three main morphs recognized (red, silver and black), as well as cross-breeding variations of these. Red is most common in Australia. Throat and abdomen in most individuals are white, as is tip of bushy tail. Ears and lower legs are black. **DISTRIBUTION** First introduced into Australia in Vic in 1850s and now widespread through all but tropical north. **HABITAT AND HABITS** Occurs in a variety of habitat types, including desert, forests, woodland, grassland, agricultural areas and urban environments. Feeds on a range of foods, although mainly carnivorous. Hunts birds, reptiles, mammals and insects, but also feeds on eggs, carrion and human rubbish, such as fruits and vegetables. Breeding occurs from June to October. Litter sizes vary subject to food abundance and age of female, but typically contain up to six pups.

■ Eared Seals ■

Otariidae, Eared Seals

Long-nosed Fur-seal ■ *Arctocephalus forsteri* TL 1.5–2.5m
(New Zealand Fur-seal)

DESCRIPTION Thickset with long, narrow, pointed snout. Brown to blackish-grey, flecked with silvery-white, paler on chest and flanks, and snout light greyish or brownish with long, pale grey whiskers. Male has longer fur (mane) on neck, heavily streaked with silvery-grey. **DISTRIBUTION** Southern coast and islands from south-western WA to southern Qld and Tas. Breeds on islands along southern WA, southern SA, south-western Vic and southern Tas coasts. **HABITAT AND HABITS** Occurs in coastal marine areas. Hunts for fish, squid, crustaceans and penguins, and basks on rocks, preferring assemblages of boulders rather than open platforms.

Female with pup   *Adult male*

Australian Fur-seal ■ *Arctocephalus pusillus* TL 1.36–2.27m
(Cape Fur Seal)

DESCRIPTION Adult males much larger than females and young males, grey-brown (darker with age), with long, silvery mane of coarse hair on neck and shoulders. Females more slender, silvery-grey above and brownish below, becoming more yellowish on chest and throat. **DISTRIBUTION** Coast and islands along southern Australia, and oceans bounded by continental shelf. Mainly Vic, Tas and islands of Bass Strait, but also extending to NSW and SA. **HABITAT AND HABITS** Hauls out and breeds on rocky areas along coast. Forages in open oceans to a depth of about 150m, for fish, squid and octopuses (and occasionally seabirds). Moves to more open rocky platforms in colonies shared with Long-nosed Fur-seal (above).

▪ Eared Seals ▪

Subantarctic Fur-seal ▪ *Atrocephalus tropicalis* TL 1.19–1.8m

DESCRIPTION Adult male larger than female. Rich brown to greyish-black above, with yellowish-cream on face and chest. Neck short and thick, with thick mane forming erectable crest on head. Female and young male paler and lack crest (which starts to

develop as male matures). **DISTRIBUTION** In Australian zone, breeds on Macquarie Island (and rarely on Heard Island), with individuals occasionally visiting beaches of Tas, southern Australian mainland and offshore islands. **HABITAT AND HABITS** Found in oceans and on islands. Semi-aquatic, feeding on fish, squid, octopuses and occasionally penguins.

Australian Sea-lion ▪ *Neophoca cinerea* TL 1.3–2.25m

DESCRIPTION Adult male larger than female, with thickened neck and chest. Coat (pelage) brown to blackish (becoming darker with age), and paler whitish-brown on head and nape. Adult female and young male silvery-grey above and yellowish-cream

below. Snout blunt. **DISTRIBUTION** Southern Australia, breeding on about 70 islands and parts of mainland along coast of SA (mainly) and WA. **HABITAT AND HABITS** Breeding and haul-out sites include sandy beaches, rocky islands, reefs and vegetated dunes. Forages at sea for variety of prey, including fish, squid, sharks, crustaceans and seabirds.

▪ EARED SEALS/EARLESS SEALS ▪

New Zealand Sea-lion ▪ *Phocarctos hookeri* TL 1.6–2.7m

DESCRIPTION Adult male much larger than female, dark brown to blackish-grey, with defined mane on head and shoulders. Adult female generally paler silvery-grey after moult, becoming more brownish during year, but some individuals can be more yellowish. **DISTRIBUTION** Mainly subantarctic islands and oceans south of New Zealand. Also on Macquarie Island in Australian waters. **HABITAT AND HABITS** Occurs on beaches and adjacent vegetated areas on islands, and surrounding marine areas. Feeds on squid, fish, octopuses and crustaceans, and occasionally on penguins and seal pups.

PHOCIDAE, EARLESS SEALS

Leopard Seal ▪ *Hydrurga leptonyx* TL 2.5–3.8m

DESCRIPTION Dark bluish-grey above, darker on head and neck, and paler silvery-white below, with numerous darker and paler grey spots. Head large, with rounded snout and powerful jaws. **DISTRIBUTION** Southern Ocean, with irregular sightings (mainly young seals) on Tas, NSW, Southern Qld, Vic, SA and WA. **HABITAT AND HABITS** Found in Antarctic and subantarctic waters. Preference for hauling out on ice floes, where it is mainly solitary, feeding on penguins (and other seabirds), krill, fish, squid and young seals.

EARLESS SEALS/CATS

Southern Elephant Seal ■ *Mirounga leonina* TL 3.5–5.8m

DESCRIPTION Male is the largest seal and is unmistakable, with enlarged, trunk-like nose that is inflated when displaying. Female much smaller with pug-like face. Generally dark grey when wet and dark reddish-brown when dry. DISTRIBUTION Infrequent visitor to NSW, Vic, Tas (breeding recorded), SA and WA. HABITAT AND HABITS Occurs on sandy or

pebbly beaches with easy access to and from water. Feeds on squid, octopuses and fish, which are mainly hunted in colder waters.

Male *Female*

FELIDAE, CATS

Domestic Cat
■ *Felis catus* TL 610–930mm, including tail 230–330mm

DESCRIPTION Highly variable, from white to black with any manner of tones of yellow-red and brown between. Most individuals are mottled or striped. DISTRIBUTION Introduced from Europe, and found in all states and territories. HABITAT AND HABITS

Cats are superbly designed predators that are adaptable to all Australian conditions. This adaptability, along with lack of naturally occurring predators in Australia, has contributed to devastation it causes to wildlife. Essentially nocturnal, cats rest during day in a den, usually a burrow, small cave or hole in a tree. Opportunistic hunter, diet largely influenced by habitat.

HORSES & ASSES/PIGS

EQUIDAE, HORSES AND ASSES

Horse ■ *Equus caballus* TL c.3m, including tail c.0.8m

DESCRIPTION Usually brown to black or mixture of both, sometimes with white flashes on feet and around muzzle. Domesticated horses occasionally breed with feral horses, producing many other colour combinations. **Donkey** *E. asinus*, introduced in Australia and found in scattered areas, has whitish muzzle, eyering and underparts, and comparatively longer ears. **DISTRIBUTION** Introduced. Found in scattered populations around Australia. **HABITAT AND HABITS** Occurs mainly in open habitats, such as open woodland and grassland. Mainly eats grass but will occasionally browse on shrubs and fruits. Requires fresh water at least every few days, and has been noted digging in dry creek beds to obtain water. Usually lives in small groups of up to ten. Responsible for widespread environmental degradation and ecosystem destruction, dramatically affecting native animal population sizes and distribution, vegetation and stream morphology.

SUIDAE, PIGS

Pig
■ *Sus scrofa* TL 1.3–1.9m, including tail 200–300mm

DESCRIPTION Varied, from white to black with any manner of tones of yellow-red and brown between. Most adults are black or dark brown, juveniles are usually mottled or striped. **DISTRIBUTION** Introduced. Occurs over most of eastern Australia and parts of northern Australia, including Top End NT and Kimberley region WA. **HABITAT AND HABITS** Found across most habitats, prefers wetter areas, and can be active both day and night. Digs out wallows in moist ground to form mud baths. True omnivores, feeding on small animals, carrion, tubers, vegetation, fruits and succulents. Feeding and wallowing behaviour has caused siginficant damage to many ecosystems across Australia.

CAMELIDAE, CAMELS AND RELATIVES

One-humped Camel
■ *Camelus dromedarius* TL 2.2–4m, including tail up to 70cm

DESCRIPTION Entirely grey-brown or predominantly pale grey-brown on body with darker hump, shoulder and back of neck. Winter coat longer and darker, shed in early summer. Single hump consisting of tissues and stored fat varies in size and shape depending on animal's condition. **DISTRIBUTION** First introduced from Canary Islands in 1840s. Since recorded in all mainland states except ACT, but most numerous in arid inland. **HABITAT AND HABITS** Strongly associated with desert environment, where preferred diet is succulent herbage, high in water and, often, salt content. Grasses make up between 20–40 per cent of diet (70–90 per cent in cattle). Also eats fruit, leaves and stems of many

shrubs and will badly damage some native trees such as the Quandong or Native Peach *Santalum acuminatum*. Long legs and neck enable it to browse trees much further from the ground than any other terrestrial herbivore in Australia. May go for long periods without drinking, however in summer drinks every day if possible. Tends to spend nights on areas of bare ground away from waterhole, watering at first light and then moving off.

BOVIDAE, CATTLE, SHEEP AND GOATS

Swamp Buffalo
■ *Bubalus bubalis* TL 2.7–3.7m, including tail up to 70cm

DESCRIPTION Dark grey to blackish, legs and hindquarters may be flushed with brown. Large horns on both sexes. **DISTRIBUTION** Introduced, initially on Melville Island and later Port Essington. Wild populations confined to northern NT. **HABITAT AND**

HABITS Lives in flooded grassland and woodland. Active day and night, sheltering away from heat of day in wallows or shade of trees. Mainly eats aquatic vegetation, but in dry season, when food is less abundant, will graze on other grasses. Solitary for most of year, but occasionally lives in small, same-sex groups. Mating season lasts for 8 months, peaking around March. Causes significant environmental degradation and is a carrier of bovine tuberculosis.

Cattle, Sheep & Goats

Goat ■ *Capra hircus* TL 1.2–1.75m, including tail 120–170mm

DESCRIPTION Typically short-haired, but fur may be longer around head and neck on males, and around hindquarters of either sex. Both sexes grow horns, but longer and more heavily-built in males. Male's beard extends down throat and onto chest. **DISTRIBUTION** Introduced since early European settlement, mainly as a source of meat and milk for settlers and to support the growing fibre industry. Since spread across 3 per cent of the continent, including many offshore islands, but most numerous in south-east. **HABITAT AND HABITS** Prefers areas with low woody vegetation. Breeding can take place year-round in arid areas or within a six month period (peaking around February) in temperate regions.

Female

Male

Chital Deer
■ *Axis axis* TL 1.5–2.1m, including tail c.25cm

DESCRIPTION Coat reddish to chestnut brown with dark stripe down back from nape to tip of tail. Muzzle dark brown to black. White spots from hindeck, along back and sides to rump and thighs. White on abdomen, rump, throat, insides of legs and ears, and underside of tail. Upper throat bright white. Tail noticeably long. Antlers, present only on bucks, have three tines. Similar looking Fallow Deer (see p. 148), has a white heart-shaped patch on rump, bordered by black, and heavier antlers. **DISTRIBUTION** Introduced and recorded from eastern states, predominantly below 1,000m asl. **HABITAT AND HABITS** Prefers woodland, forests and clearings near permanent waterways. Feeds on grasses, supplemented with leaves, stems, fruits, seeds, flowers and bark.

CATTLE, SHEEP & GOATS

Hog Deer ◼ *Axis porcinus*
TL 1.1–1.5m, including tail 120–140mm

DESCRIPTION Smallest of deer species in Australia, roughly the size of a sheep. Coat ranges from dark brown to rich reddish-brown, greyer on legs, and some individuals have darker line along back. Underparts darker greyish-brown. Back slopes upwards to high rump. Tail tipped white. Antlers short, normally with three tines, but can have more. **DISTRIBUTION** Introduced. Confined to coastal plains and islands of southern and eastern Vic. **HABITAT AND HABITS** Inhabits damp woodland, heathland, scrubland, fringes of cleared grassland and coastal dunes. Feeds mainly on grasses, sedges, herbs and, less commonly, foliage of shrubs. Mostly nocturnal.

Red Deer
◼ *Cervus elaphus* TL 1.8–2.65mm, including tail 120–150mm

DESCRIPTION Second largest of Australia's deer species (females much smaller than males). Coat red-brown above with cream underbelly. Rump patch cream and extends onto back. Adult male antlers U-shaped, with five to eight tines on each; upper tines in cluster and lowest two forward facing and close together. Tail very short. Similar Sambar Deer (opposite) is darker with simpler antlers (maximum of 3 tines on each). **DISTRIBUTION** Introduced. Established on south-east Australian mainland, from south-eastern Qld through NSW, ACT and Vic to eastern SA. **HABITAT AND HABITS** Grazes and browses on a range of plant matter, lichen and fungi. Forms small same-sex herds for most of year, males disperse in late March and April to attract and breed with groups of females.

CATTLE, SHEEP & GOATS

Rusa Deer
Cervus timorensis TL 1.7–2.05m, including tail 180–250mm

DESCRIPTION Coat coarse, grey to greyish-brown. Chest and throat paler. Darker line runs down chest and between forelegs. Long tufts of light hair extend from inner ears. Very vocal compared to closely related Sambar Deer (below) and ears less rounded. Antlers have three tines, rear one noticeably longer and almost vertical at tip. Old antlers shed around December and regrown following June. **DISTRIBUTION** Introduced and widely distributed on Australian mainland. Recorded on islands of northern NT and Torres Strait, but most numerous between south-eastern Qld to eastern Vic and south-eastern SA. **HABITAT AND HABITS** Occurs in open areas, and grazes on soft grasses, but will also browse on coarser plant material. Crepuscular and nocturnal. Feeds on grasses, but will also browse on other vegetation. Gregarious in single-sex herds outside breeding times. Males decorate antlers with vegetation when displaying.

Sambar Deer
Cervus unicolor TL 1.85–2.8m, including tail 250–300mm

DESCRIPTION Largest of Australia's deer. Coat uniform light or dark brown, greyish or black, occasionally pale buff under chin, between forelegs and under body. Ears prominent and bat-like, with pale inner. Antlers with three tines. Old ones shed around November and news ones fully grown by April. Rusa Deer (above) has pale lips and obviously longer rear antler tine. **DISTRIBUTION** Introduced to Vic in 1860s and now widespread through central and eastern Vic, ACT and eastern NSW (most numerous in south-east). Also introduced to Coburg Peninsula NT. **HABITAT AND HABITS** Occurs in forests and woodland, grazes on leaves and other coarse vegetation. Typically crepuscular and nocturnal, but can be seen during day and easily startled. Stags generally solitary while hinds and young form small groups.

Cattle, Sheep & Goats/Pygmy Right Whale

Fallow Deer
■ *Dama dama* TL 1.6–1.95m, including tail 200–240mm

DESCRIPTION Medium-sized deer with long tail. Coat can be black, white, grey-brown (common) or light-brown. Rump patch is white, heart-shaped and bordered with black or brown. Buck has prominent 'Adam's apple'. Antlers have many tines and become flattened in older individuals. Old antlers are shed around October and regrown following March. Similar looking Chital Deer (see p. 145) lacks blackish border around rump patch and has striking white upper throat. **DISTRIBUTION** Introduced and widely distributed. Mostly found in south-east, from south-east Qld to southern SA, and Tas. **HABITAT AND HABITS** Generally associated with fringes of clearings, where it grazes on grasses and herbs, but also browses on coarser material. Typically crepuscular and nocturnal, but can be active during day if disturbed.

Neobalaenidae, Pygmy Right Whale

Pygmy Right Whale ■ *Caperea marginata* TL up to 6.5m

DESCRIPTION Smallest baleen whale. Dark bluish-grey above and whitish below between mouth and tail, with paler diagonal stripe between blowhole and belly, and darker eye-patch. Head long and narrow with arched jaw. Dorsal fin sickle shaped and angled backwards (absent in Southern Right Whale, see opposite). **DISTRIBUTION** Southern hemisphere, with most sightings in Australia from strandings in WA, SA, Vic, Tas and south-eastern NSW. **HABITAT AND HABITS** Found in temperate and subantarctic oceans. Travels alone or in small groups, and feeds on plankton, which is sieved through baleen from mouthfuls of water skimmed from surface of the ocean.

Stranded indiviual on Tasmanian beach

RIGHT WHALE/RORQUALS

BALAENIDAE, RIGHT WHALE

Southern Right Whale
■ *Eubalaena australis* TL 13.5–17m

DESCRIPTION Large, thickset whale (weighing up to 80 tonnes), with no dorsal fin and body tapering to thin tail. Generally dark brown to bluish-black, with numerous white callosities on head and around jaw. Blow distinctly V shaped. **DISTRIBUTION** In Australia, most commonly seen from south-western WA to south-eastern NSW. **HABITAT AND HABITS** Occurs in temperate and subantarctic seas, and coastal waters in Australia, where it is often seen close to beaches and headlands. Spends the summer months feeding on krill in Antarctic waters, and calves in coastal areas of Australia in May–October.

BALAENOPTERIDAE, RORQUALS

Sei Whale ■ *Balaenoptera borealis* TL 12–21m

DESCRIPTION Dark bluish-grey above and mottled grey below, with white, pleated throat. Body long and thin, with flat head, angular lower jaw and rostrum with central ridge. Dorsal fin angled backwards and located two-thirds along body, flippers small and tail flattened laterally. Blow about 3m high and cone shaped. **DISTRIBUTION** Worldwide from polar regions to tropics, but only uncommonly seen in Australian waters. **HABITAT AND HABITS** Found throughout open oceans. Travels either alone or in pairs, and feeds on planktonic crustaceans, including krill, and small shoaling fish by skimming the surface or lunging sideways at the surface.

RORQUALS

Bryde's Whale ■ *Balaenoptera brydei* TL 15.6m

DESCRIPTION Similar to Sei Whale (see p. 149). Dark brownish-grey above, with 3 ridges on rostrum and hook-shaped dorsal fin, and white below, with throat pleats extending to or past navel. **DISTRIBUTION** Circumglobal, but rarely sighted above 35 degrees South.

HABITAT AND HABITS Occurs in tropical, subtropical and temperate oceans. Feeds year round on small schooling fish and some crustaceans and cephalopods. Two 'forms' recognized, with larger form inhabiting deeper oceans and smaller form more inshore. Thought to be species complex. Travels either alone or in small groups.

Blue Whale ■ *Balaenoptera musculus* TL 21–33m

DESCRIPTION Largest of the world's living mammals. Bluish-grey above with mottled paler blotches that assist identification, and paler below with white throat. Prominent ridge around blowhole, merging with ridge on rostrum. Dorsal fin appears small and located well back on body. **DISTRIBUTION** Recorded from all states in Australian region. Two recognized subspecies in Australian waters. The Antarctic Blue Whale *B. m. intermedia*, the largest (>30m), is an infrequent visitor. The smaller (about 24m) Pygmy Blue Whale *B. m. brevicauda* is more common. **HABITAT AND HABITS** Migratory. Lives in deeper and, to a lesser extent, shallower oceans. Lunge feeds, singly or in pairs, on krill and other crustaceans.

◾ RORQUALS ◾

Fin Whale ◾ *Balaenoptera physalus* TL 25m

DESCRIPTION Grey to greyish-brown above, with right-hand side of lower jaw whitish, and white below, including underside and front edges of flippers. Rostrum has central ridge. Dorsal fin tall and about 75 per cent along body from snout. **DISTRIBUTION** Found in all the world's oceans. **HABITAT AND HABITS** Found in deeper oceans. Usually travels in pairs or small groups, but larger groups assemble to feed on high concentrations of krill. Also feeds on other crustaceans, fish and squid.

Dwarf Minke Whale ◾ *Balaenoptera acutorostrata* TL 7.8m

DESCRIPTION Smallest baleen whale. Dark blue-grey above, with narrow, pointed rostrum that has single ridge, and white underparts, extending higher onto sides in places. Underside of tail white with black patch, and flippers with white band near base and remainder blue-grey, which readily identifies it from very similar (larger) Antarctic Minke Whale (see p. 152). **DISTRIBUTION** In Australian region, recorded in all states except NT. **HABITAT AND HABITS** Found in coastal waters. Mostly solitary, feeding on krill and small shoaling fish, which are often herded together before it lunges through the condensed shoal. Baleen plates used to filter out prey from water.

▪ Rorquals ▪

Antarctic Minke Whale ▪ *Balaenoptera bonaerensis* TL 9.4m

DESCRIPTION Dark blue-grey above, with narrow, pointed rostrum with single ridge. Paler greyish patch on sides behind flipper, and white below. Underside of flipper and tail white, and more white parts in baleen on right side. May be confused with Dwarf Minke Whale (see p. 151). **DISTRIBUTION** Broadly distributed in southern hemisphere. Antarctic waters in summer, and winters in lower latitudes. **HABITAT AND HABITS** Mainly oceanic, with preference for colder (icy) waters during summer months. Feeds almost continuously on krill, either singly or in small groups of up to 8, in areas of pack ice.

Humpback Whale ▪ *Megaptera novaeangliae* TL 16m

DESCRIPTION Greyish-black above and sharply contrasting white below. Flippers long and slender (up to 5m) with knobs with barnacles on leading edge. Tail flukes have black-and-white markings that are different in each individual. **DISTRIBUTION** Within Australian waters, most numerous along east and west coasts. **HABITAT AND HABITS** Found in coastal waters, as it travels south in summer to feed on krill in Antarctic waters, and returns northwards to winter calving and breeding areas in warmer subtropical and tropical waters. Well known for dramatic breaches, often rolling or spinning in the air and landing horizontally.

◾ Sperm Whale/Pygmy & Dwarf Sperm Whales ◾

Physeteridae, Sperm Whale

Sperm Whale ◾ *Physeter macrocephalus* TL 11–18m

DESCRIPTION Generally dark grey, with whitish patches on body (particularly on throat and snout), which become more extensive with age. Head large and square, with blowhole at front, and mouth long and thin. Dorsal fin reduced to small, triangular lump. **DISTRIBUTION** Found in all oceans around Australia, and throughout the rest of the world. **HABITAT AND HABITS** Occurs in deeper waters (>600m), often in herds of about 30 individuals, where it dives to great depths to forage on or close to the sea floor for medium to large squid, and similarly sized sharks, rays and octopuses.

Kogiidae, Pygmy and Dwarf Sperm Whales

Pygmy Sperm Whale ◾ *Kogia breviceps* TL 3.6m

DESCRIPTION Dark greyish-brown above and pale below, tinged with pink. Head large, blunt and rounded, extending past narrow jaw, and dorsal fin small and hook shaped. Flippers set high up on sides. **DISTRIBUTION** Temperate, subtropical and tropical waters. Within Australian region, strandings recorded in all states. **HABITAT AND HABITS** Found in deep waters beyond continental shelf, either singly or in small groups of up to 6 individuals. Feeds on cephalopods, fish and crustaceans.

DOLPHINS

DELPHINIDAE, DOLPHINS

Short-beaked Common Dolphin ■ *Delphinus delphis* TL 1.7–2.5m

DESCRIPTION Dark brown to greyish on back, merging with yellowish-cream flank, forming hourglass pattern. Tail usually light grey, and belly with occasional spotting.

Dorsal fin brown to light grey, slightly sickle shaped. Beak long, narrow and predominantly dark. Narrow stripe from eye to flipper. **DISTRIBUTION** Known from Atlantic and Pacific Oceans. **HABITAT AND HABITS** Found in shallow coastal to deep oceanic waters. Sighted in groups of 10 to 1,000 individuals. Diet consists of small schooling fish and squid.

Melon-headed Whale ■ *Peponocephala electra* TL up to 2.8m

DESCRIPTION Dark grey to dark brown, sometimes with blackish cape, more blackish on face, and with variable paler grey patterning on throat and chest. Head slim, slightly bulbous and somewhat conical, appearing triangular when viewed from above or below. Dorsal fin tall angled backwards, midway along back. At sea, Melon-headed Whale can be difficult to distinguish from **Pygmy Killer Whale** *Feresa attenuata*, which shares the same range. Melon-headed has pointed flippers, Pygmy Killer has rounded flippers and a more bulbous head. False Killer Whale (see p. 157) is significantly larger and flippers have distinct, wide elbow. **DISTRIBUTION** Tropical and subtropical seas, south to around 35°S. **HABITS AND HABITATS** Prefers deep, warm waters, seldom inshore. Feeds largely on small fish and squid. Forms large groups of up to around 500 individuals. Strong swimmer with some 'porpoising' and will bow-ride boats.

154

■ Dolphins ■

Short-finned Pilot Whale ■ *Globicephala macrorhynchus*
Long-finned Pilot Whale ■ *G. melas* TL 5.5–7m

DESCRIPTION Dark brownish to greyish-black above. Dorsal fin rounded with long base, with white (*melas*) or greyish (*macrorhynchus*) patch on back, and white patches on underside. Flippers long, with distinct elbow, and much longer in *melas* than in *macrorhynchus*.
DISTRIBUTION *Macrorhynchus* found in tropical to temperate waters; *melas* favours more temperate to subpolar waters.
HABITAT AND HABITS Oceanic and coastal. Generally gregarious in groups up to 100 individuals or more. Feeds mainly on squid, but also eats octopuses and fish.

Risso's Dolphin ■ *Grampus griseus* TL 3–4m

DESCRIPTION Greyish, with conspicuous scarring along head and body, becoming almost white with age. Dorsal fin in middle of back, and side flippers long and thin. Head blunt and rounded (no beak), with diagnostic longitudinal ridge on forehead. **DISTRIBUTION** Temperate to tropical waters and over continental shelf, with sightings from all Australian states. **HABITAT AND HABITS** Found in deep water (>400m) near continental shelf. Groups of about 30 and occasionally up to several thousand individuals feed mainly on cephalopods and some krill.

◾ Dolphins ◾

Dusky Dolphin ◾ *Lagenorhynchus obscurus* TL 1.6–2.1m

DESCRIPTION Dark grey to blackish above, paler on top of tail, and greyish on flanks, with darker colours extending onto silvery-white underparts. Pale greyish-white, crescent-shaped markings extend along tail and sides. **DISTRIBUTION** Wide ranging in southern hemisphere. Records from southern oceans of Australia, mainly from Tas, but also WA, SA and Vic. **HABITAT AND HABITS** Found in deeper waters (<2,000m). Normally forms groups of about 20 individuals, but occasionally up to 1,000 or more. Feeds on small fish, squid and krill.

Killer Whale ◾ *Orcinus orca* TL 9.6m
(Orca)

DESCRIPTION Black body with white or yellowish white patch behind eye, and greyish saddle behind fin. Underside white, extending onto flanks behind each fin. Snout blunt and dorsal fin tall. **DISTRIBUTION** Recorded in all coastal and oceanic marine regions (including rivers and estuaries) worldwide. **HABITAT AND HABITS** Found in coastal waters and deeper oceans. Travels in groups, usually of up to 10 individuals, but occasionally up to 50 (more rarely several hundred). Groups usually led by oldest female, and hunt as a team, showing high level of cooperation. Feeds on huge variety of prey, including seals, seabirds, fish, squid and other cetaceans.

DOLPHINS

False Killer Whale ■ *Pseudorca crassidens* TL 5.3m

DESCRIPTION Dark grey to black with pale grey chest-patch, and long, blunt head with rounded snout and long jaw. Dorsal fin tall, rounded and hook shaped, located midway along body. Flippers have distinctive elbow. **DISTRIBUTION** Major oceans worldwide. Subtropical to tropical waters off Australia. **HABITAT AND HABITS** Generally found in deeper oceans past the continental shelf, but also in shallower inlets. Forms herds usually of 10–50 individuals, occasionally up to 800, which feed on fish, squid and some dolphins and young whales.

Australian Hump-backed Dolphin ■ *Sousa sahulensis* TL 2.6m

DESCRIPTION Greyish above, paler and washed with pink in older individuals, and whitish on flanks and belly, often with some darker spotting. Beak long and cylindrical, and dorsal fin short and triangular, with slightly curved rear edge and wide base. **DISTRIBUTION** Shallow inshore waters of northern Australia, from central WA to central NSW. **HABITAT AND HABITS** Found in coastal waters, generally shallower than 20m and within 5km of coast, including harbours, estuaries, rivers and shallow reefs. Feeds on fish, cephalopods and some crustaceans.

◼ Dolphins ◼

Spinner Dolphin ◼ *Stenella longirostris* TL 2m

DESCRIPTION Dark grey above, blacker on cape and dorsal fin, and paler towards the head. Whitish below, darker near the tail. Beak long and slender, with a darker greyish stripe passing through eye to the greyish-black flippers. **DISTRIBUTION** Northern

Australia, between Bunbury, south-western WA, and north-eastern NSW. **HABITAT AND HABITS** Occurs in deeper tropical and subtropical waters. Feeds mainly at night on small fish, squid and prawns, within large pods of several hundred individuals. Spins longitudinally during aerial jumps.

Striped Dolphin ◼ *Stenella coeruleoalba* TL 2.6m

DESCRIPTION Grey above, more bluish-grey from forehead to upper tail, and with blackish line along flanks and under tail. Rest of underparts white. Dark grey line between flipper and eye. Beak moderately short. **DISTRIBUTION** From around Albany WA, through northern Australia to around Eden NSW. **HABITAT AND HABITS** Found in deeper tropical and subtropical oceans. Cathemeral, feeding on small fish, squid and prawns. Often seen in groups of 10–20, but can form pods of several hundred to several thousand individuals.

DOLPHINS

Rough-toothed Dolphin ■ *Steno bredanensis* TL 2.6m

DESCRIPTION Dark cape and paler greyish body above, with pale to pink belly, occasionally more blotched, and darker from rear of vent and onto flattened tail. Long, cone-shaped beak merges with forehead. **DISTRIBUTION** Tropical and subtropical seas of northern and eastern Australia, from around Barrow Island north-western WA to southern NSW. **HABITAT AND HABITS** Found in warmer deep oceans, in groups of up to 20 individuals. Feeds on pelagic fish (such as Dolphin Fish) and cephalopods, staying submerged for up to 15 minutes while foraging.

Indo-pacific Bottle-nosed Dolphin ■ *Tursiops aduncus*
Common Bottle-nosed Dolphin ■ *T. truncatus* TL 2.4–3.1m
(*truncatus* larger than *aduncus*)

DESCRIPTION Grey to brownish-grey above, paler below. Beak with upwardly curving mouth, and dorsal fin backwardly sloping and moderately tall. Beak longer and flippers larger in *T. aduncus* than in *T. truncatus*. **DISTRIBUTION** Temperate and tropical waters worldwide. Found in all Australian states and NT, but more common in southern oceans. **HABITAT AND HABITS** Occurs in coastal waters (*aduncus*) and deeper oceans (*truncatus*), but can be found in other areas, making identification very difficult. Both species feed mainly on fish and cephalopods. Capable of forming very large pods of tens of thousands, but usual groups number around 30 individuals. Previously considered synonymous, and much debate on size and extent of populations.

Indo-pacific Bottle-nosed Dolphin

Checklist of the Mammals of Australia

AUSTRALIAN MAMMAL LIST

Common and scientific names follow Jackson & Groves, 2015, and Australian Bat Society (see Reference section for more information).

For each species, an 'x' indicates presence in a particular state or territory; a '-' indicates past presence of extinct species.

State abbreviations:
Q	Queensland
NS	New South Wales
A	Australian Capital Territory
V	Victoria
T	Tasmania
S	South Australia
W	Western Australia
NT	Northern Territory
AAT	Australian Antarctic Territory and Antarctic Islands

Abbreviations of IUCN Red List status:
EX	Extinct
CR	Critically Endangered
EN	Endangered
VU	Vulnerable
NT	Near Threatened
LC	Least Concern
DD	Data Deficient
NE	Not Evaluated

Common Name	Scientific Name	Q	NS	A	V	T	S	W	NT	AAT	IUCN
Ornithorhynchidae (Platypus)											
Platypus	*Ornithorhynchus anatinus*	x	x	x	x	x	x				NT
Tachyglossidae (Echidnas)											
Short-beaked Echidna	*Tachyglossus aculeatus*	x	x	x	x	x	x	x	x		LC
Western Long-beaked Echidna++	*Zaglossus bruijni*							-			CR
Dasyuridae (Dasyurids)											
Brush-tailed Mulgara	*Dasycercus blythi*	x						x	x		LC
Crest-tailed Mulgara	*Dasycercus cristicauda*	x					x		x		NT
Kaluta	*Dasykaluta rosamondae*							x			LC
Kowari	*Dasyuroides byrnei*	x					x				VU
Western Quoll	*Dasyurus geoffroyi*							x			NT
Northern Quoll	*Dasyurus hallucatus*	x						x	x		EN
Spotted-tailed Quoll	*Dasyurus maculatus*	x	x	x	x	x					NT
Eastern Quoll	*Dasyurus viverrinus*					x					EN
Dibbler	*Parantechinus apicalis*							x			EN

Common Name	Scientific Name	Q	NS	A	V	T	S	W	NT	AAT	IUCN
Sandstone Pseudantechinus	Pseudantechinus bilarni								x		LC
Fat-tailed Pseudantechinus	Pseudantechinus macdonnellensis						x	x	x		LC
Carpentarian Pseudantechinus	Pseudantechinus mimulus	x							x		NT
Ningbing Pseudantechinus	Pseudantechinus ningbing							x	x		LC
Tan Pseudantechinus	Pseudantechinus roryi							x			LC
Woolley's Pseudantechinus	Pseudantechinus woolleyae							x			LC
Tasmanian Devil	Sarcophilus harrisii					x					EN
Rusty Antechinus	Antechinus adustus	x									LC
Agile Antechinus	Antechinus agilis		x	x	x						LC
Silver-headed Antechinus	Antechinus argentus	x									NE
Black-tailed Antechinus	Antechinus arktos	x	x								NE
Fawn Antechinus	Antechinus bellus								x		VU
Yellow-footed Antechinus	Antechinus flavipes	x	x	x	x		x	x			LC
Atherton Antechinus	Antechinus godmani	x									LC
Cinnamon Antechinus	Antechinus leo	x									LC
Mainland Dusky Antechinus	Antechinus mimetes		x	x	x						NE
Swamp Antechinus	Antechinus minimus				x	x	x				LC
Buff-footed Antechinus	Antechinus mysticus	x									NE
Brown Antechinus	Antechinus stuartii	x	x	x							LC
Subtropical Antechinus	Antechinus subtropicus	x	x								LC
Tasmanian Dusky Antechinus	Antechinus swainsonii					x					LC
Tasman Peninsula Dusky Antechinus	Antechinus vandycki					x					NE
Red-tailed Phascogale	Phascogale calura							x			NT
Northern Phascogale	Phascogale pirata								x		VU
Brush-tailed Phascogale	Phascogale tapoatafa	x	x	x	x		x	x			NT
Gile's Planigale	Planigale gilesi	x	x		x		x		x		LC
Long-tailed Planigale	Planigale ingrami	x						x	x		LC
Common Planigale	Planigale maculata	x	x					x	x		LC
Narrow-nosed Planigale	Planigale tenuirostris	x	x				x		x		LC
Kultarr	Antechinomys laniger	x	x				x	x	x		LC
Wongai Ningaui	Ningaui ridei	x					x	x	x		LC
Pilbara Ningaui	Ningaui timealeyi							x			LC
Southern Ningaui	Ningaui yvonnae		x		x		x		x		LC
Chestnut Dunnart	Sminthopsis archeri	x									DD
Kakadu Dunnart	Sminthopsis bindi								x		NT
Butler's Dunnart	Sminthopsis butleri							x	x		VU
Fat-tailed Dunnart	Sminthopsis crassicaudata	x	x	x	x		x	x	x		LC
Little Long-tailed Dunnart	Sminthopsis dolichura						x	x			LC
Julia Creek Dunnart	Sminthopsis douglasi	x									NT
Grey-bellied Dunnart	Sminthopsis fuliginosus						x	x			DD
Gilbert's Dunnart	Sminthopsis gilberti						x	x			LC
White-tailed Dunnart	Sminthopsis granulipes							x			LC
Greater Hairy-footed Dunnart	Sminthopsis hirtipes						x	x	x		LC
White-footed Dunnart	Sminthopsis leucopus	x	x	x	x	x					LC
Large Long-tailed Dunnart	Sminthopsis longicaudata						x	x	x		LC

Common Name	Scientific Name	Q	NS	A	V	T	S	W	NT	AAT	IUCN
Stripe-faced Dunnart	Sminthopsis macroura	x	x				x	x	x		LC
Common Dunnart	Sminthopsis murina	x	x	x	x		x				LC
Ooldea Dunnart	Sminthopsis ooldea						x	x	x		LC
Sandhill Dunnart	Sminthopsis psammophila						x	x			VU
Red-cheeked Dunnart	Sminthopsis virginiae	x						x	x		LC
Lesser Hairy-footed Dunnart	Sminthopsis youngsoni	x						x	x		LC
Myrmecobalidae (Numbat)											
Numbat	Myrmecobius fasciatus							x			EN
Thylacinidae (Thylacine)											
Thylacine++	Thylacinus cyanocephalus					-					EX
Notoryctidae (Marsupial Moles)											
Northern Marsupial Mole	Notoryctes caurinus							x			LC
Southern Marsupial Mole	Notoryctes typhlops						x	x	x		LC
Chaeropodidae (Pig-footed Bandicoot)											
Southern Pig-footed Bandicoot++	Chaeropus ecaudatus		-		-		-	-			EX
Northern Pig-footed Bandicoot++	Chaeropus yirratji sp. nov.						-	-	-		EX
Peramelidae (Bandicoots)											
Long-nosed Echymipera	Echymipera rufescens	x									LC
Golden Bandicoot	Isoodon auratus							x	x		VU
Quenda	Isoodon fusciventer							x			NE
Northern Brown Bandicoot	Isoodon macrourus	x	x					x	x		LC
Southern Brown Bandicoot	Isoodon obesulus		x	x	x	x	x				LC
Cape York Brown Bandicoot	Isoodon peninsulae	x									NE
Shark Bay (Western Barred) Bandicoot	Perameles bougainville							x	x		VU
Desert Bandicoot++	Perameles eremiana		-				-	-	-		EX
Liverpool Plains Striped Bandicoot++	Perameles fasciata		-								NE
Eastern Barred Bandicoot	Perameles gunnii				x	x					VU
Marl++	Perameles myosuros						-	-			NE
Southern Long-nosed Bandicoot	Perameles nasuta	x	x	x	x						LC
South-eastern Striped Bandicoot++	Perameles notina		-		-		-				NE
Northern Long-nosed Bandicoot	Perameles pallescens	x									NE
Nullarbor Barred Bandicoot++	Perameles papillon sp. nov.						-	-			NE
Thylacomyidae (Bilbies)											
Greater Bilby	Macrotis lagotis	x						x	x	x	VU
Lesser Bilby++	Macrotis leucura						-	-	-		EX
Phascolarctidae (Koala)											
Koala	Phascolarctos cinereus	x	x	x	x		x				VU
Vombatiidae (Wombats)											
Northern Hairy-nosed Wombat	Lasiorhinus krefftii	x									CR
Southern Hairy-nosed Wombat	Lasiorhinus latifrons		x		x		x	x			NT
Bare-nosed (Common) Wombat	Vombatus ursinus		x	x	x	x	x				LC
Burramyidae (Pygmy-possums)											
Mountain Pygmy-possum	Burramys parvus		x		x						CR

Common Name	Scientific Name	Q	NS	A	V	T	S	W	NT	AAT	IUCN
Long-tailed Pygmy Possum	Cercartetus caudatus	x									LC
Western Pygmy-possum	Cercartetus concinnus				x		x	x			LC
Little Pygmy-possum	Cercartetus lepidus				x	x	x				LC
Eastern Pygmy-possum	Cercartetus nanus	x	x	x	x	x	x				LC
Petauridae (Striped Possum, Leadbeater's Possum and Lesser Gliders)											
Torresian Striped Possum	Dactylopsila trivirgata	x									LC
Leadbeater's Possum	Gymnobelideus leadbeateri				x						CR
Savanna Glider	Petaurus ariel	x						x	x		NE
Yellow-bellied Glider	Petaurus australis	x	x	x	x		x				NT
Sugar Glider	Petaurus breviceps	x	x	?							LC
Mahogany Glider	Petaurus gracilis	x									EN
Krefft's Glider	Petaurus notatus	x	x	x	x	x	x				NE
Squirrel Glider	Petaurus norfolcensis	x	x	x	x						LC
Pseudocheridae (Ring-tailed Possums and Greater Gliders)											
Lemuroid Ring-tailed Possum	Hemibelideus lemuroides	x									NT
Central Greater Glider	Petauroides armillatus	x									NE
Northern Greater Glider	Petauroides minor	x									NE
Southern Greater Glider	Petauroides volans	x	x	x	x						VU
Western Ring-tailed Possum	Pseudocheirus occidentalis							x			CR
Eastern Ring-tailed Possum	Pseudocheirus peregrinus	x	x	x	x	x	x				LC
Daintree River Ring-tailed Possum	Pseudochirulus cinereus	x									NT
Herbert River Ring-tailed Possum	Pseudochirulus herbertensis	x									LC
Rock Ring-tailed Possum	Petropseudes dahlii	x						x	x		LC
Green Ring-tailed Possum	Pseudochirops archeri	x									NT
Tarsipedidae (Honey Possum)											
Honey Possum	Tarsipes rostratus							x			LC
Acrobatidae (Feather-tailed Gliders)											
Broad-toed Feather-tailed Glider	Acrobates frontalis	x	x	x	x		x				NE
Narrow-toed Feather-tailed Glider	Acrobates pygmaeus	x	x	x	x		x				LC
Phalangeridae (Cuscuses and Brush-tailed Possums)											
Southern Common Cuscus	Phalanger mimicus	x									LC
Australian Spotted Cuscus	Spilocuscus nudicaudatus	x									LC
Short-eared Brush-tailed Possum	Trichosurus caninus	x	x	x							LC
Mountain Brush-tailed Possum	Trichosurus cunninghami		x	x	x						LC
Common Brush-tailed Possum	Trichosurus vulpecula	x	x	x	x	x	x	x	x		LC
Scaly-tailed Possum	Wyulda squamicaudata							x	x		NT
Hypsiprymnodontidae (Musky Rat-Kangaroo)											
Musky Rat-kangaroo	Hypsiprymnodon moschatus	x									LC
Potoroidae (Bettongs and Potoroos)											
Rufous Bettong	Aepyprymnus rufescens	x	x								LC
Desert Bettong++	Bettongia anhydra							-	-		EX
Eastern Bettong	Bettongia gaimardi					x					NT
Burrowing Bettong	Bettongia lesueur		x					x	x		NT
Brush-tailed Bettong	Bettongia penicillata							x			CR

Common Name	Scientific Name	Q	NS	A	V	T	S	W	NT	AAT	IUCN
Nullabor Dwarf Bettong++	Bettonga pusilla						-	-			EX
Northern Bettong	Bettonga tropica	x									EN
Desert Rat-kangaroo++	Caloprymnus campestris	-					-				EX
Gilbert's Potoroo	Potorous gilbertii							x			CR
Long-footed Potoroo	Potorous longipes		x		x						VU
Broad-faced Potoroo++	Potorous platyops						-	-			EX
Long-nosed Potoroo	Potorous tridactylus	x	x	x	x	x	x				NT
Macropodidae (Kangaroos and Wallabies)											
Bennett's Tree-kangaroo	Dendrolagus bennettianus	x									NT
Lumholtz's Tree-kangaroo	Dendrolagus lumholtzi	x									NT
Allied Rock-wallaby	Petrogale assimilis	x									LC
Western Short-eared Rock-wallaby	Petrogale brachyotis							x	x		LC
Monjon Rock-wallaby	Petrogale burbidgei							x			NT
Cape York Rock-wallaby	Petrogale coenensis	x									EN
Nabarlek Rock-wallaby	Petrogale concinna							x	x		EN
Godman's Rock-wallaby	Petrogale godmani	x									NT
Herbert's Rock-wallaby	Petrogale herberti	x									LC
Unadorned Rock-wallaby	Petrogale inornata	x									LC
Black-footed Rock-wallaby	Petrogale lateralis						x	x	x		VU
Mareeba Rock-wallaby	Petrogale mareeba	x									NT
Brush-tailed Rock-wallaby	Petrogale penicillata	x	x	x	x						VU
Proserpine Rock-wallaby	Petrogale persephone	x									EN
Purple-necked Rock-wallaby	Petrogale purpureicollis	x									NT
Rothschild's Rock-wallaby	Petrogale rothschildi							x			LC
Mount Claro Rock-wallaby	Petrogale sharmani	x									VU
Eastern Short-eared Rock-wallaby	Petrogale wilkinsini								x		NE
Yellow-footed Rock-wallaby	Petrogale xanthopus	x	x				x				NT
Rufous-bellied Pademelon	Thylogale billardierii					x					LC
Red-legged Pademelon	Thylogale stigmatica	x	x								LC
Red-necked Pademelon	Thylogale thetis	x	x								LC
Central Hare-wallaby++	Lagorchestes asomatus							-	-		EX
Spectacled Hare-wallaby	Lagorchestes conspicillatus	x						x	x		LC
Rufous Hare-wallaby	Lagorchestes hirsutus		x					x	x		VU
Eastern Hare-wallaby++	Lagorchestes leporides		-		-		-				EX
Western Grey Kangaroo	Macropus fuliginosus	x	x		x		x	x			LC
Eastern Grey Kangaroo	Macropus giganteus	x	x	x	x	x	x				LC
Agile Wallaby	Notamacropus agilis	x						x	x		LC
Black-striped Wallaby	Notamacropus dorsalis	x	x								LC
Tammar Wallaby	Notamacropus eugenii						x	x			LC
Toolache Wallaby++	Notamacropus greyi				-		-				EX
Western Brush Wallaby	Notamacropus irma							x			LC
Parma Wallaby	Notamacropus parma	x	x								NT
Whiptail Wallaby	Notamacropus parryi	x	x								LC
Red-necked Wallaby	Notamacropus rufogriseus	x	x	x	x	x					LC
Bridled Nailtail Wallaby	Onychogalea fraenata	x	x								VU

Common Name	Scientific Name	Q	NS	A	V	T	S	W	NT	AAT	IUCN
Crescent Nailtail Wallaby++	Onychogalea lunata		-				-	-	-		EX
Northern Nailtail Wallaby	Onychogalea unguifera	x						x	x		LC
Antilopine Wallaby	Osphranter antilopinus	x						x	x		LC
Black Wallaroo	Osphranter bernardus								x		NT
Common Wallaroo	Osphranter robustus	x	x	x	x		x	x	x		LC
Red Kangaroo	Osphranter rufus	x	x	x	x		x	x	x		LC
Swamp Wallaby	Wallabia bicolor	x	x	x	x		x				LC
Quokka	Setonix brachyurus							x			VU
Banded Hare-wallaby	Lagostrophus fasciatus							x			VU
Dugongidae (Dugong)											
Dugong	Dugong dugon	x						x	x		VU
Muridae (Rats and Mice)											
Capricorn Rabbit-rat++	Conilurus capricornensis	-									EX
White-footed Tree-rat++	Conilurus albipes	-	-			-		-			EX
Brush-tailed Rabbit-rat	Conilurus penicillatus	x						x	x		VU
Water Rat	Hydromys chrysogaster	x	x	x	x	x	x	x	x		LC
Forrest's Mouse	Leggadina forresti	x	x				x	x	x		LC
Northern Short-tailed Mouse	Leggadina lakedownensis	x						x	x		LC
Lesser Stick-nest Rat++	Leporillus apicalis		-				-	-	-		EX
Greater Stick-nest Rat	Leporillus conditor		x				x	x			NT
Broad-toothed Rat	Mastacomys fuscus		x	x	x	x					NT
Grassland Melomys	Melomys burtoni	x	x					x	x		LC
Cape York Melomys	Melomys capensis	x									LC
Fawn-footed Melomys	Melomys cervinipes	x	x								LC
Bramble Cay Melomys++	Melomys rubicola	-									EX
Black-footed Tree-rat	Mesembriomys gouldii	x							x		VU
Golden-backed Tree-rat	Mesembriomys macrurus							x	x		NT
Spinifex Hopping-mouse	Notomys alexis	x					x	x	x		LC
Short-tailed Hopping-mouse++	Notomys amplus						-	-	-		EX
Northern Hopping-mouse	Notomys aquilo								x		EN
Fawn Hopping-mouse	Notomys cervinus	x	x				x		x		NT
Dusky Hopping-mouse	Notomys fuscus	x	x				x		x		VU
Long-tailed Hopping Mouse++	Notomys longicaudatus		-					-	-		EX
Big-eared Hopping-mouse++	Notomys macrotis							-			EX
Mitchell's Hopping-mouse	Notomys mitchellii		x		x		x	x			LC
Darling Downs Hopping-mouse++	Notomys mordax	-									EX
Broad-cheeked Hopping-mouse++	Notomys robustus						-				EX
Ash-grey Mouse	Pseudomys albocinereus							x			LC
Silky Mouse	Pseudomys apodemoides				x		x				LC
Long-eared Mouse++	Pseudomys auritus				-		-				EX
Plains Mouse	Pseudomys australis	x					x		x		VU
Bolam's Mouse	Pseudomys bolami		x				x	x			LC
Kakadu Pebble-mouse	Pseudomys calabyi								x		VU
Western Pebble-mouse	Pseudomys chapmani							x			LC
Delicate Mouse	Pseudomys delicatulus	x						x	x		LC

Common Name	Scientific Name	Q	NS	A	V	T	S	W	NT	AAT	IUCN
Desert Mouse	Pseudomys desertor	x					x	x	x		LC
Shark Bay Mouse#	Pseudomys fieldi							x			VU
Smoky Mouse	Pseudomys fumeus		x	x	x						VU
Blue-grey Mouse++	Pseudomys glaucus	-	-								EX
Gould's Mouse++	Pseudomys gouldii		-				-	-			EX
Eastern Chestnut Mouse	Pseudomys gracilicaudatus	x	x								LC
Sandy Inland Mouse	Pseudomys hermannsburgensis	x	x				x	x	x		LC
Long-tailed Mouse	Pseudomys higginsi					x					LC
Central Pebble-mouse	Pseudomys johnsoni	x						x	x		LC
Western Chestnut Mouse	Pseudomys nanus	x						x	x		LC
New Holland Mouse	Pseudomys novaehollandiae	x	x		x	x					VU
Western Mouse	Pseudomys occidentalis							x			NT
Hastings River Mouse	Pseudomys oralis	x	x								VU
Eastern Pebble-mouse	Pseudomys patrius	x									LC
Heath Mouse	Pseudomys shortridgei				x		x	x			NT
Giant White-tailed Rat	Uromys caudimaculatus	x									LC
Pygmy White-tailed Rat	Uromys hadrourus	x									NT
Water Mouse	Xeromys myoides	x							x		VU
Common Rock-rat	Zyzomys argurus	x						x	x		LC
Arnhem Land Rock-rat	Zyzomys maini								x		VU
Carpentarian Rock-rat	Zyzomys palatilis								x		CR
Central Rock-rat	Zyzomys pedunculatus								x		CR
Kimberley Rock-rat	Zyzomys woodwardi							x			LC
House Mouse+	Mus musculus	x	x	x	x	x	x	x	x		LC
Dusky Rat	Rattus colletti								x		LC
Pacific Rat+	Rattus exulans		x					x		x	LC
Bush Rat	Rattus fuscipes	x	x	x	x		x	x			LC
Cape York Rat	Rattus leucopus	x									LC
Swamp Rat	Rattus lutreolus	x	x	x	x	x	x				LC
Maclear's Rat++	Rattus macleari							-			EX
Bulldog Rat++	Rattus nativitatis							-			EX
Brown Rat+	Rattus norvegicus	x	x	x	x	x	x	x	x	x	LC
Black Rat+	Rattus rattus	x	x	x	x	x	x	x	x	x	LC
Canefield Rat	Rattus sordidus	x							x		LC
Oriental House Rat+	Rattus tanezumi							?			LC
Pale Field Rat	Rattus tunneyi	x						x	x		LC
Long-haired Rat	Rattus villosissimus	x					x	x	x		LC
Sciuridae (Squirrels)											
Northern Palm Squirrel+	Funambulus pennantii							x			LC
Leporidae (Hares and Rabbits)											
European Brown Hare+	Lepus europaeus	x	x	x	x	x	x				LC
European Rabbit+	Oryctolagus cuniculus	x	x	x	x	x	x	x	x		EN
Soricidae (Shrews)											
Christmas Island Shrew	Crocidura trichura							x			CR
Pteropodidae (Old World Fruit Bats)											
Northern Blossom-bat	Macroglossus minimus	x						x	x		LC
Eastern Blossom-bat	Syconycteris australis	x	x								LC

Common Name	Scientific Name	Q	NS	A	V	T	S	W	NT	AAT	IUCN
Eastern Tube-nosed Bat	Nyctimene robinsoni	x	x								LC
Bare-backed Flying-fox	Dobsonia magna	x									LC
Black Flying-fox	Pteropus alecto	x	x	x	x			x	x		LC
Percy Island Flying-fox++	Pteropus brunneus	-									EX
Spectacled Flying-fox	Pteropus conspicillatus	x									EN
Large-eared Flying-fox	Pteropus macrotis	?									LC
Christmas Island Flying-fox	Pteropus natalis							x			NE
Grey-headed Flying-fox	Pteropus poliocephalus	x	x	x	x		x				VU
Little Red Flying-fox	Pteropus scapulatus	x	x	x	x			x	x		LC
Megadermatidae (Ghost Bat)											
Ghost Bat	Macroderma gigas	x						x	x		VU
Rhinolophidae (Horseshoe Bats)											
Eastern Horseshoe-Bat	Rhinolophus megaphyllus	x	x	x	x						LC
Large-eared Horseshoe-bat	Rhinolophus robertsi	x									LC
Hipposideridae (Leaf-nosed Bats)											
Dusky Leaf-nosed Bat	Hipposideros ater	x						x	x		LC
Fawn Leaf-nosed Bat	Hipposideros cervinus	x									LC
Diadem Leaf-nosed Bat	Hipposideros diadema	x									LC
Arnhem Leaf-nosed Bat	Hipposideros inornatus								x		VU
Semon's Leaf-nosed Bat	Hipposideros semoni	x									LC
Northern Leaf-nosed Bat	Hipposideros stenotis	x						x	x		LC
Rhinonicteris (Orange Leaf-nosed Bat)											
Orange Leaf-nosed Bat	Rhinonicteris aurantia	x						x	x		LC
Emballonuridae (Sheath-tailed Bats)											
Yellow-bellied Sheath-tailed Bat	Saccolaimus flaviventris	x	x	x	x		x	x	x		LC
Cape York Sheath-tailed Bat	Saccolaimus mixtus	x									NT
Bare-rumped Sheath-tailed Bat	Saccolaimus saccolaimus	x							x		LC
Coastal Sheath-tailed Bat	Taphozous australis	x									NT
Common Sheath-tailed Bat	Taphozous georgianus	x						x	x		LC
Hill's Sheath-tailed Bat	Taphozous hilli	x					x	x	x		LC
Arnhem Sheath-tailed Bat	Taphozous kapalgensis							x	x		LC
Troughton's Sheath-tailed Bat	Taphozous troughtoni	x									LC
Molossidae (Free-tailed Bats)											
White-striped Free-tailed Bat	Austronomus australis	x	x	x	x	x	x	x	x		LC
Great Northern Free-tailed Bat	Chaerephon jobensis	x						x	x		LC
Eastern Coastal Free-tailed Bat	Micronomus norfolkensis	x	x	x	x						NT
Northern Coastal Free-tailed Bat	Ozimops cobourgianus							x	x		LC
Cape York Free-tailed Bat	Ozimops halli	x									DD
Western Free-tailed Bat	Ozimops kitcheneri							x			LC
Northern Free-tailed Bat	Ozimops lumsdenae	x						x	x		LC
Inland Free-tailed Bat	Ozimops petersi	x	x	x	x		x	x	x		LC
Southern Free-tailed Bat	Ozimops planiceps		x	x	x		x	x			LC
Ride's Free-tailed Bat	Ozimops ridei	x	x	x	x		x				LC
Bristle-faced Free-tailed Bat	Setirostris eleryi	x	x				x		x		NT
Miniopteridae (Bent-winged Bats)											
Little Bent-winged Bat	Miniopterus australis	x	x								LC
Large Bent-winged Bat	Miniopterus orianae	x	x	x	x		x	x	x		NE

Common Name	Scientific Name	Q	NS	A	V	T	S	W	NT	AAT	IUCN
Vespertilionidae (Vespertilionid Bats)											
Golden-tipped Bat	Phoniscus papuensis	x	x								LC
Flute-nosed Bat	Murina florium	x									LC
Arnhem Long-eared Bat	Nyctophilus arnhemensis	x						x	x		LC
Eastern Long-eared Bat	Nyctophilus bifax	x	x								LC
Corben's Long-eared Bat	Nyctophilus corbeni	x	x		x		x				VU
Pallid Long-eared Bat	Nyctophilus daedalus	x						x	x		LC
Lesser Long-eared Bat	Nyctophilus geoffroyi	x	x	x	x	x	x	x	x		LC
Gould's Long-eared Bat	Nyctophilus gouldi	x	x	x	x		x				LC
Holt's Long-eared Bat	Nyctophilus holtorum sp. nov							x			NE
Lord Howe Long-eared Bat++	Nyctophilus howensis		-								EX
Greater Long-eared Bat	Nyctophilus major							x			LC
Tasmanian Long-eared Bat	Nyctophilus sherrini					x					VU
Pygmy Long-eared Bat	Nyctophilus walkeri	x						x	x		LC
Large-eared Wattled Bat	Chalinolobus dwyeri	x	x								VU
Gould's Wattled Bat	Chalinolobus gouldii	x	x	x	x	x	x	x	x		LC
Chocolate Wattled Bat	Chalinolobus morio	x	x	x	x	x	x	x	x		LC
Hoary Wattled Bat	Chalinolobus nigrogriseus	x	x					x	x		LC
Little Pied Wattled Bat	Chalinolobus picatus	x	x		x		x				NT
Western Falsistrelle	Falsistrellus mackenziei							x			NT
Eastern Falsistrelle	Falsistrellus tasmaniensis	x	x	x	x	x	x				LC
Forest Pipistrelle	Pipistrellus adamsi	x							x		LC
Christmas Island Pipistrelle++	Pipistrellus murrayi							-			EX
Northern Pipistrelle	Pipistrellus westralis	x						x	x		LC
Greater Broad-nosed Bat	Scoteanax rueppellii	x	x	x	x						LC
Inland Broad-nosed Bat	Scotorepens balstoni	x	x	x	x		x	x	x		LC
Little Broad-nosed Bat	Scotorepens greyii	x	x	x	x		x	x	x		LC
Eastern Broad-nosed Bat	Scotorepens orion	x	x	x	x						LC
Northern Broad-nosed Bat	Scotorepens sanborni	x						x	x		LC
Central-eastern Broad-nosed Bat	Scotorepens sp. (Parnaby)	x	x								NE
Inland Forest-bat	Vespadelus baverstocki	x	x		x		x	x	x		LC
Northern Cave-bat	Vespadelus caurinus	x						x	x		LC
Large Forest-bat	Vespadelus darlingtoni	x	x	x	x	x	x				LC
Yellow-lipped Cave-bat	Vespadelus douglasorum							x			LC
Finlayson's Cave-bat	Vespadelus finlaysoni	x					x	x	x		LC
Eastern Forest-bat	Vespadelus pumilus	x	x								LC
Southern Forest-bat	Vespadelus regulus	x	x	x	x	x	x	x			LC
Eastern Cave-bat	Vespadelus troughtoni	x	x								LC
Little Forest-bat	Vespadelus vulturnus	x	x	x	x	x	x				LC
Large-footed Myotis	Myotis macropus	x	x	x	x		x	x	x		LC
Canidae (Dogs)											
Dingo+	Canis familiaris	x	x	x	x	x	x	x	x		NE
Red Fox+	Vulpes vulpes	x	x	x	x	x	x	x	x		LC
Otariidae (Eared Seals)											
Long-nosed Fur-seal	Arctocephalus forsteri	x	x		x	x	x	x		x	LC
Antarctic Fur-seal	Arctocephalus gazella									x	LC
Australian Fur Seal	Arctocephalus pusillus		x		x	x	x			x	LC

Common Name	Scientific Name	Q	NS	A	V	T	S	W	NT	AAT	IUCN
Subantarctic Fur-seal	Arctocephalus tropicalis				x					x	LC
Australian Sea-lion	Neophoca cinerea				x		x	x			EN
New Zealand Sea-lion	Phocarctos hookeri									x	EN
Phocidae (Earless Seals)											
Leopard Seal	Hydrurga leptonyx	x	x		x	x	x	x		x	LC
Weddell Seal	Leptonychotes weddellii									x	LC
Crabeater Seal	Lobodon carcinophaga									x	LC
Southern Elephant Seal	Mirounga leonina		x		x	x	x	x		x	LC
Ross Seal	Ommatophoca rossii									x	LC
Mustelidae (Weasels, Badgers, Skunks and Otters)											
Euopean Polecat (Ferret)+	Mustela putorius	x				x		x			LC
Felidae (Cats)											
Domestic Cat+	Felis catus	x	x	x	x	x	x	x	x		NE
Equidae (Horses and Asses)											
Donkey+	Equus asinus	x					x	x	x		NE
Horse+	Equus caballus	x	x	x	x			x	x	x	NE
Suidae (Pigs)											
Pig+	Sus scrofa	x	x	x	x	x	x	x	x		LC
Camelidae (Camels and Relatives)											
One-humped Camel+	Camelus dromedarius	x	x				x	x	x		NE
Bovidae (Cattle, Sheep and Goats)											
Banteng+	Bos javanicus								x		EN
Cattle+	Bos taurus	x	x	x	x	x	x	x	x		NE
Swamp Buffalo+	Bubalus bubalis								x		NE
Blackbuck+	Antilope cervicapra							?			LC
Sheep+	Capra aries	x	x	x	x	x	x	x	x		NE
Goat+	Capra hircus	x	x	x	x	x	x	x	x		NE
Ceridae (Deer)											
Chital Deer+	Axis axis	x	x		x		x				LC
Hog Deer+	Axis porcinus				x						EN
Red Deer+	Cervus elaphus	x	x	x	x		x				LC
Rusa Deer+	Cervus timorensis	x	x	x	x		x		x		VU
Sambar Deer+	Cervus unicolor		x	x	x						VU
Fallow Deer+	Dama dama	x	x	x	x	x	x	x			LC
Neobalaenidae (Pygmy Right Whale)											
Pygmy Right Whale	Caperea marginata		x		x	x	x	x		x	LC
Balaenidae (Right Whales)											
Southern Right Whale	Eubalaena australis	x	x		x	x	x	x		x	LC
Balaenopteridae (Roquals)											
Dwarf Minke Whale	Balaenoptera acutorostrata	x	x		x	x	x	x			LC
Antarctic Minke Whale	Balaenoptera bonaerensis	x	x		x	x	x	x		x	NT
Sei Whale	Balaenoptera borealis	x	x		x	x	x	x	x	x	EN
Bryde's Whale+++	Balaenoptera brydei	x	x		x	x	x	x	x		NE
Eden's Whale	Balaenoptera edeni	x	x		x	x	x	x	x		LC
Blue Whale	Balaenoptera musculus	x	x		x	x	x	x	x	x	EN
Omura's Whale	Balaenoptera omurai	x						x	x		DD
Fin Whale	Balaenoptera physalus	x	x		x	x	x	x	x	x	VU
Humpback Whale	Megaptera novaeangliae	x	x		x	x	x	x	x	x	LC

Common Name	Scientific Name	Q	NS	A	V	T	S	W	NT	AAT	IUCN
Physeteridae (Sperm Whales)											
Sperm Whale	*Physeter macrocephalus*	x	x		x	x	x	x	x	x	VU
Kogiidae (Pygmy and Dwarf Sperm Whales)											
Pygmy Sperm Whale	*Kogia breviceps*	x	x		x	x	x	x	x		LC
Dwarf Sperm Whale	*Kogia sima*	x	x		x	x	x	x	x		LC
Ziphiidae (Beaked Whales)											
Arnoux's Beaked Whale	*Berardius arnuxii*		x		x	x	x	x		x	LC
Southern Bottle-nosed Whale	*Hyperodon planifrons*		x		x	x	x	x		x	LC
Longman's Beaked Whale	*Indopacetus pacificus*	x						x			LC
Andrew's Beaked Whale	*Mesoplodon bowdoini*		x		x	x	x	x		x	DD
Blainville's Beaked Whale	*Mesoplodon densirostris*	x	x		x	x	x	x	x		LC
Gingko-toothed Beaked Whale	*Mesoplodon gingkodens*	x	x		x				x		DD
Gray's Beaked Whale	*Mesoplodon grayi*		x		x	x	x	x		x	LC
Hector's Beaked Whale	*Mesoplodon hectori*		x		x	x	x	x		x	DD
Strap-toothed Beaked Whale	*Mesoplodon layardii*	x	x		x	x	x	x		x	LC
True's Beaked Whale	*Mesoplodon mirus*				x	x	x	x		x	LC
Tasman Beaked Whale	*Tasmacetus shepherdi*				x	x	x	x		x	DD
Cuvier's Beaked Whale	*Ziphius cavirostris*	x	x		x	x	x	x		x	LC
Delphinidae (Dolphins and Killer Whales)											
Short-beaked Common Dolphin	*Delphinus delphis*	x	x		x	x	x	x	x		LC
Pygmy Killer Whale	*Feresa attenuata*	x	x					x			LC
Short-finned Pilot Whale	*Globicephala macrorhynchus*	x	x		x		x	x	x		LC
Long-finned Pilot Whale	*Globicephala melas*	x	x		x	x	x	x	x		LC
Risso's Dolphin	*Grampus griseus*	x	x		x	x	x	x	x		LC
Fraser's Dolphin	*Lagenodelphis hosei*	x	x		x			x	x		LC
Hourglass Dolphin	*Lagenorhynchus cruciger*									x	LC
Dusky Dolphin	*Lagenorhynchus obscurus*				x	x	x	x			LC
Southern Rightwhale Dolphin	*Lissodelphis peronii*				x	x	x	x			LC
Australian Snub-finned Dolphin	*Orcaella heinsohni*	x						x	x		VU
Killer Whale	*Orcinus orca*	x	x		x	x	x	x	x	x	DD
Melon-headed Whale	*Peponocephala electra*	x	x					x	x		LC
False Killer Whale	*Pseudorca crassidens*	x	x		x	x	x	x	x	x	NT
Australian Hump-backed Dolphin	*Sousa sahulensis*	x	x					x	x		VU
Pantropical Spotted Dolphin	*Stenella attenuata*	x	x					x	x		LC
Striped Dolphin	*Stenella coeruleoalba*	x	x		x			x	x		LC
Spinner Dolphin	*Stenella longirostris*	x	x					x	x		LC
Rough-toothed Dolphin	*Steno bredanensis*	x	x		x			x	x		LC
Indo-pacific Bottle-nosed Dolphin	*Tursiops aduncus*	x	x		x	x	x	x	x		NT
Common Bottle-nosed Dolphin	*Tursiops truncatus*	x	x		x	x	x	x	x		LC
Phocoenidae (Porpoises)											
Spectacled Porpoise	*Phocoena dioptrica*					x				x	LC

+Introduced
++Extinct within Australia
+++Taxon still to be formally recognized
Recent evidence suggests that the extant Shark Bay Mouse is conspecific with the previously considered extinct Gould's Mouse.

FURTHER INFORMATION

Websites
Australian Bat Society www.ausbats.org.au
Australian Mammal Society www.australianmammals.org.au
Australian Museum www.australianmuseum.net.au
Australian Wildlife Conservancy www.australianwildlife.org
Department of Environment and Energy www.environment.gov.au
Kape Images www.kapeimages.com.au
Museum and Art Gallery of the Northern Territory www.magnt.net.au
Museums Victoria www.museumvictoria.com.au
Perth Museum www.museum.wa.gov.au
Peter Rowland Photographer & Writer www.prpw.com.au
Queensland Museum www.qm.qld.gov.au
South Australian Museum www.samuseum.sa.gov.au
Tasmanian Museum and Art Gallery www.tmag.tas.gov.au
Wildlife Tourism Australia www.wildlifetourism.org.au
WWF Australia www.wwf.org.au

References
Andrew, D. (2015). *The Complete Guide to Finding the Mammals of Australia*. CSIRO Publishing, Clayton South, Victoria.
Baker, A. M., Mutton, T. Y. & Hines, H. B. (2013). A new dasyurid marsupial from Kroombit Tops, south-east Queensland, Australia: the Silver-headed Antechinus, *Antechinus argentus* sp. nov. (Marsupialia: Dasyuridae). *Zootaxa* 3746 (2): 201–209.
Baker, A. M., Mutton, T. Y., Hines, H. B. & Van Dyck, S. (2014). The Black-tailed Antechinus, *Antechinus arktos* sp. nov.: a new species of carnivorous marsupial from montane regions of the Tweed Volcano caldera, eastern Australia. *Zootaxa* 3765 (2): 101–133.
Baker, A. M., Mutton, T. Y., Mason, E. D. & Gray, E. L. (2015). A Taxonomic Assessment of the Australian Dusky Antechinus Complex; a New Species, the Tasman Peninsula Dusky Antechinus (*Antechinus vandycki* sp. nov.) and an Elevation to Species of the Mainland Dusky Antechinus (*Antechinus swainsonii* mimetes (Thomas)). *Memoirs of the Queensland Museum – Nature* 59: 75–126. Brisbane.
Baker, A. M. & Van Dyck, S. (2015). Taxonomy and redescription of the Swamp Antechinus, *Antechinus minimus* (È. Geoffroy) (Marsupialia: Dasyuridae). *Memoirs of the Queensland Museum – Nature* 59: 127–170. Brisbane.
Churchill, S. (2008). *Australian Bats*. 2nd edn. Allen & Unwin, Crows Nest, Sydney.
Cremona, T., Baker, A. M., Cooper, S. J. B, Montague-Drake, R., Stobo-Wilson, A.M., & Carthew, S.M. (2020) Integrative taxonomic investigation of *Petaurus breviceps* (Marsupialia: Petauridae) reveals three distinct species. *Zoological Journal of the Linnean Society*, 2021, 191: 503–527.
Jackson, S. & Groves, C. (2015). *Taxonomy of Australian Mammals*. CSIRO Publishing, Clayton South, Victoria.
Jackson, S. M., Groves, C.P., Fleming, P. J., Aplin, K. P., Eldridge, M. D., Gonzalez, A. & Helgen, K. M. (2017). The wayward dog: Is the Australian native dog or dingo a distinct species? *Zootaxa*, 4317 (2), 201–224.
Jedensjö, M., Kemper, C. M., Milella, M., Willems, E. P. & Krützen (2020) Taxonomy and distribution of bottlenose dolphins (genus *Tursiops*) in Australian waters: an osteological clarification *Canadian Journal of Zoology*. 98: 461–479.
Mason, Eugene D., Burwell, Chris J., & Baker, Andrew M. (2015). Prey of the silver-headed antechinus (*Antechinus argentus*), a new species of Australian dasyurid marsupial. *Australian Mammalogy* 37(2): 164–169.
Menkhorst, P. & Knight, F. (2011). *A Field Guide to the Mammals of Australia*. 3rd edn. Allen & Unwin, Crows Nest, Sydney.
Parnaby, H. E. (2009). A taxonomic review of Australian Greater Long-eared Bats previously known as *Nyctophilus timoriensis* (Chiroptera: Vespertilionidae) and some associated taxa. *Australian Zoologist*

■ ACKNOWLEDGEMENTS ■

vol. 35 (1): 39–81.
Parnaby, H. E., King, A. G., Eldridge, M. D. B. (2021). A new bat species from southwestern Western Australia, previously assigned to Gould's Long-eared Bat *Nyctophilus gouldi* Tomes, 1858. *Records of the Australian Museum* 73 (1): 53–66.
Potter, Sally, Close, Robert, Taggart, David A. & Eldridge, Mark (2014). Taxonomy of rock-wallabies, *Petrogale* (Marsupialia: Macropodidae). IV. Multifaceted study of the brachyotis group identifies additional taxa. *Australian Journal of Zoology* 62 (5) 401–414.
Reardon, T. B., McKenzie, N. L., Cooper, S. J. B., Appleton, B., Carthew, S. & Adams, M. (2014). A molecular and morphological investigation of species boundaries and phylogenetic relationships in Australian free-tailed bats *Mormopterus* (Chiroptera: Molossidae). *Australian Journal of Zoology* 2014, 62: 109–136.
Shirihai, Hardoran & Jarrett, Brett (2006). *Whales, Dolphins and Seals Field Guide to Marine Mammals of the World*. A & C Black Publishers Ltd, London.
Stobo-Wilson, A.M. (2018) Ecology of the savanna glider (*Petaurus ariel*) in tropical northern Australia. College of Engineering, IT and Environment, Charles Darwin University (Submitted in fulfillment of the requirements for the Degree of Doctor of Philosophy [BSc Hons]).
Travouillon K. J., Phillips M. J. (2018). Total evidence analysis of the phylogenetic relationships of bandicoots and bilbies (Marsupialia: Peramelemorphia): reassessment of two species and description of a new species. *Zootaxa* 4378 (2), 224–256.
Tavouillon, K. J., Simões, B. F., Miguez, R. P., Brace, S., Brewer, P., Stemmer, D., Price, G. J., Cramb, J. & Louy, J. (2019). Hidden in plain sight: reassessment of the Pig-footed Bandicoot, *Chaeropus ecaudatus* (Peramelemorphia, Chaeropodidae), with a description of a new species from central Australia, and use of the fossil record to trace its past distribution. *Zootaxa* 4566 (1) 1–69.
The IUCN Red List of Threatened Species. Version 2020-21. www.iucnredlist.org. Downloaded on 6 March 2017.
Van Dyck, Stephen, Gynther, Ian & Baker, Andrew (2013). *Field Companion to the Mammals of Australia*. New Holland Publishers.
Woinorski, J. C. Z. & Burbidge, A. A. (2014). *The Action Plan for Australian Mammals 2012*. CSIRO Publishing, Victoria.

ACKNOWLEDGEMENTS

The authors thank all of the people who contributed their time, advice and photographic material to this book. Special thanks go to Angus McNab for reviewing text and images, and for generously supplying a large number of the images used in the book; David Donnelly, Paul Meek (NSW DPI - Invasive Animals), Greg Ford (Balance Environmental), John Harris (Wildlife Experiences), Terry Tweedie and Brett Jarrett for reviewing sections of text and images for accuracy; Carly Starr, Bronny Read, Australian Geographic, Julie Chaise (WWF Australia), Sharon Wormleaton, Tim Bawden, Lisa Nunn (Ninox Photography), Jim Shields, Michael Hitchcock (CVA) and Anders Zimny for providing additional information, images and support; and Thomas Rowland for providing the illustrations. The many photographers who have supplied us with their wonderful images most readily, we thank you for your generosity and assistance. Our family, friends and colleagues for their encouragement and unreserved support, both during the undertaking of this book and in all of our times of need. The researchers and authors of published material that we have used to gather information and produce this publication, without their enthusiasm and passion for the mammals of Australia this kind of book would not be possible. The people who administer and review the accuracy and legitimacy of information on the web-based databases that were consulted for taxonomic revisions, geographical mapping and links to further resources, and, of course, the people who supply their field sightings and research information to these sites. The access to this information was invaluable in many instances, and the willingness of the broad mammalogy community as a whole, which unselfishly make comprehensive data available to the public, can never be overstated. Lastly, and by no means least, we thank the publishers, their wonderful staff, particularly John Beaufoy and Rosemary Wilkinson, for the opportunity to produce this book, Krystyna Mayer for her thoroughness in reviewing and editing the text, and assisting with the overall book design, and Sally Bird of Calidris Literary Agency for her long-standing support.

INDEX

Acrobates frontalis 62
　pygmaeus 62
Aepyprymnus rufescens 65
Antechinomys laniger 35
Antechinus, Agile 28
　Black-tailed 29
　Brown 32
　Buff-footed 31
　Dusky Mainland 30
　Dusky Tasmanian 32
　Fawn 29
　Rusty 28
　Swamp 31
　Yellow-footed 30
Antechinus adustus 28
　agilis 28
　arktos 29
　bellus 29
　flavipes 30
　mimetes 30
　minimus 31
　mysticus 31
　stuartii 32
　swainsonii 32
Arctocephalus forsteri 139
　pusillus 139
　tropicalis 140
Austronomus australis 120
Axis axis 145
　porcinus 146

Balaenoptera acutorostrata 151
　bonaerensis 152
　borealis 149
　brydei 150
　musculus 150
　physalus 151
Bandicoot, Brown Northern 45
　Brown Southern 46
　Eastern Barred 47
　Northern Long-nosed 48
　Shark Bay 46
　Southern Long-nosed 47
Bat, Arnhem Long-eared 124
　Bare-rumped Sheath-tailed 118
　Bristle-faced Free-tailed 122
　Chocolate Wattled 129
　Coastal Sheath-tailed 118
　Common Sheath-tailed 119

Corben's Long-eared 125
Diadem Leaf-nosed 116
Dusky Leaf-nosed 115
Eastern Broad-nosed 132
Eastern Long-eared 125
Eastern Tube-nosed 111
Ghost 114
Golden-tipped 124
Gould's Long-eared 127
Gould's Wattled 128
Greater Broad-nosed 131
Hill's Sheath-tailed 119
Hoary Wattled 129
Inland Broad-nosed 131
Large Bent-winged 123
Lesser Long-eared 126
Little Bent-winged 123
Little Broad-nosed 132
Little Pied Wattled 130
Northern Broad-nosed 133
Northern Free-tailed 121
Northern Leaf-nosed 116
Orange Leaf-nosed 117
Pallid Long-eared 126
Pygmy Long-eared 128
Ride's Free-tailed 122
Southern Free-tailed 121
Tasmanian Long-eared 127
Troughton's Sheath-tailed 120
White-striped Free-tailed 120
Yellow-bellied Sheath-tailed 117
Bettong, Brush-tailed 67
　Burrowing 66
　Eastern 66
　Northern 67
　Rufous 65
Bettongia gaimardi 66
　lesueur 66
　penicillata 67
　tropica 67
Blossom-bat, Northern 111
　Indo-pacific 159
Bubalus bubalis 144
Burramys parvus 51

Camel, One-humped 144
Camelus dromedarius 144
Canis familiaris 138
Caperea marginata 148

Capra hircus 145
Cat, Domestic 142
Cave-bat, Eastern 136
　Finlayson's 135
　Northern 134
Cercartetus caudatus 51
　concinnus 52
　lepidus 52
　nanus 53
Cervus elaphus 146
　timorensis 147
　unicolor 147
Chalinolobus gouldii 128
　morio 129
　nigrogriseus 129
　picatus 130
Cuscus, Australian Spotted 63
　Southern Common 62

Dactylopsila trivirgata 53
Dama dama 148
Dasycercus blythi 21
　cristicauda 21
Dasykaluta rosamondae 22
Dasyuroides byrnei 22
Dasyurus geoffroyi 23
　hallucatus 23
　maculatus 24
　viverrinus 24
Deer, Chital 145
　Fallow 148
　Hog 146
　Red 146
　Rusa 147
　Sambar 147
Delphinus delphis 154
Dendrolagus bennettianus 69
　lumholtzi 70
Dingo 138
Dobsonia magna 110
Dolphin, Australian Humpbacked 157
　Common Bottle-nosed 159
　Dusky 156
　Risso's 155
　Rough-toothed 159
　Short-beaked Common 154
　Spinner 158
　Striped 158
Dugong 89

INDEX

Dugong dugon 89
Dunnart, Butler's 38
  Common 42
  Fat-tailed 38
  Gilbert's 40
  Greater Hairy-footed 40
  Lesser 44
  Julia Creek 39
  Kakadu 37
  Large Long-tailed 41
  Little Long-tailed 39
  Ooldea 43
  Red-cheeked 43
  Stripe-faced 42
  White-footed 41

Echidna, Short-beaked 20
Echymipera, Long-nosed 45
Echymipera rufescens 45
Equus caballus 143
Eubalaena australis 149

Falsistrelle, Eastern 130
Falsistrellus tasmaniensis 130
Felis catus 142
Flying-fox, Black 112
  Christmas Island 113
  Grey-headed 113
  Little Red 114
  Spectacled 112
Forest-bat, Eastern 135
  Inland 133
  Large 134
  Little 137
  Southern 136
Fox, Red 138
Fruit-bat, Bare-backed 110
Fur-seal, Australian 139
  Long-nosed 139
  Subantarctic 140

Glider, Broad-toed Feather-tailed 62
  Central Greater 57 Krefft's 56
  Narrow-toed Feather-tailed 62
  Northern Greater 57
  Savanna 54
  Squirrel 56
  Southern Greater 58
  Sugar 55

  Yellow-bellied 55
Globicephala macrorhynchus 155
  melas 155
Goat 145
Grampus griseus 155
Greater Bilby 48
Gymnobelideus leadbeateri 54

Hare, European Brown 109
Hare-wallaby, Banded 88
  Rufous 79
  Spectacled 79
Hipposideros ater 115
  diadema 116
  stenotis 116
Hopping-mouse, Fawn 95
  Mitchell's 96
  Spinifex 95
Horse 143
Horseshoe-bat, Eastern 115
Hydromys chrysogaster 89
Hydrurga leptonyx 141
Hypsiprymnodon moschatus 65

Isoodon macrourus 45
  obesulus 46

Kaluta 22
Kangaroo, Eastern Grey 80
  Red 87
  Western Grey 80
Koala 49
Kogia breviceps 153
Kowari 22
Kultarr 35

Lagenorhynchus obscurus 156
Lagorchestes conspicillatus 79
  hirsutus 79
Lagostrophus fasciatus 88
Lasiorhinus krefftii 49
  latifrons 49
Leggadina forresti 90
  lakedownensis 90
Leopard Seal 141
Leporillus conditor 91
Lepus europaeus 109

Macroderma gigas 114
Macroglossus minimus 111

Macropus fuliginosus 80
  giganteus 80
Macrotis lagotis 48
Mastacomys fuscus 91
Megaptera novaeangliae 152
Melomys, Cape York 92
  Fawn-footed 93
  Grassland 92
Melomys burtoni 92
  capensis 92
  cervinipes 93
Mesembriomys gouldii 93
  macrurus 94
Miniopterus australis 123
  orianae 123
Mirounga leonina 142
Mouse, Ash-grey 96
  Delicate 97
  Desert 98
  Eastern Chestnut 99
  Forrest's 90
  Hastings River 102
  House 94
  Long-tailed 100
  New Holland 101
  Northern Short-tailed 90
  Plains 97
  Sandy Inland 99
  Smoky 98
  Western Chestnut 101
Mulgara, Crest-tailed 21
  Brush-tailed 21
Mus musculus 94
Myotis, Large-footed 137
Myotis macropus 137
Myrmecobius fasciatus 44

Neophoca cinerea 140
Ningaui, Pilbara 36
  Southern 37
  Wongai 36
Ningaui ridei 36
  timealeyi 36
  yvonnae 37
Notamacropus agilis 81
  dorsalis 81
  eugenii 82
  irma 82
  parma 83
  parryi 83

INDEX

rufogriseus 84
Notomys alexis 95
  cervinus 95
  mitchellii 96
Numbat 44
Nyctimene robinsoni 111
Nyctophilus arnhemensis 124
  bifax 125
  corbeni 125
  daedalus 126
  geoffroyi 126
  gouldi 127
  sherrini 127
  walkeri 128

Onychogalea fraenata 84
  unguifera 85
Orcinus orca 156
Ornithorhynchus anatinus 20
Oryctolagus cuniculus 110
Osphranter antilopinus 85
  bernardus 86
  robustus 86
  rufus 87
Ozimops lumsdenae 121
  planiceps 121
  ridei 122

Pademelon, Red-legged 78
  Red-necked 78
  Rufous-bellied 77
Pebble-mouse, Central 100
  Eastern 102
Peponocephala electra 154
Perameles bougainville 46
  gunnii 47
  nasuta 47
  pallescens 48
Petauroides armillatus 57
  minor 57
  volans 58
Petaurus ariel 54
  australis 55
  breviceps 55
  norfolcensis 56
  notatus 56
Petrogale assimilis 70
  brachyotis 71
  burbidgei 71
  concinna 72

  godmani 72
  herberti 73
  inornata 73
  lateralis 74
  mareeba 74
  penicillata 75
  persephone 75
  purpureicollis 76
  wilkinsi 76
  xanthopus 77
Petropseudes dahlii 60
Phalanger mimicus 62
Phascogale, Brush-tailed 33
Phascogale tapoatafa 33
Phocarctos hookeri 141
Phoniscus papuensis 124
Physeter macrocephalus 153
Pig 143
Planigale, Common 34
  Gile's 33
  Long-tailed 34
  Narrow-nosed 35
Planigale gilesi 33
  ingrami 34
  maculata 34
  tenuirostris 35
Platypus 20
Possum, Common Brush-tailed 64
  Daintree River Ring-tailed 59
  Eastern Ring-tailed 59
  Green Ring-tailed 61
  Herbert River Ring-tailed 60
  Honey 61
  Leadbeater's 54
  Mountain Brush-tailed 63
  Rock Ring-tailed 60
  Scaly-tailed 64
  Short-eared Brush-tailed 63
  Torresian Striped 53
  Western Ring-tailed 58
Potoro, Gilbert's 68
  Long-footed 68
  Long-nosed 69
Potorous gilbertii 68
  longipes 68
  tridactylus 69
Pseudantechinus, Carpentarian 26
  Fat-tailed 25

  Ningbing 26
  Sandstone 25
  Woolley's 27
Pseudantechinus bilarni 25
  macdonnellensis 25
  mimulus 26
  ningbing 26
  woolleyae 27
Pseudocheirus occidentalis 58
  peregrinus 59
Pseudochirops archeri 61
Pseudochirulus cinereus 59
  herbertensis 60
Pseudomys albocinereus 96
  australis 97
  delicatulus 97
  desertor 98
  fumeus 98
  gracilicaudatus 99
  hermannsburgensis 99
  higginsi 100
  johnsoni 100
  nanus 101
  novaehollandiae 101
  oralis 102
  patrius 102
Pseudorca crassidens 157
Pteropus alecto 112
  conspicillatus 112
  natalis 113
  poliocephalus 113
  scapulatus 114
Pygmy-possum, Eastern 53
  Little 52
  Long-tailed 51
  Mountain 51
  Western 52

Quokka 88
Quoll, Eastern 24
  Northern 23
  Spotted-tailed 24
  Western 23

Rabbit, European 110
Rat, Black 103
  Broad-toothed 91
  Brown 103
  Bush 106
  Canefield 108

INDEX

Cape York 107
Dusky 106
Giant White-tailed 104
Greater Stick-nest 91
Long-haired 109
Pale Field 108
Swamp 107
Water 89
Rat-kangaroo, Musky 65
Rattus colletti 106
  fuscipes 106
  leucopus 107
  lutreolus 107
  norvegicus 103
  rattus 103
  sordidus 108
  tunneyi 108
  villosissimus 109
Rhinolophus megaphyllus 115
Rhinonicteris aurantia 117
Rock-rat, Central 105
  Common 104
  Kimberley 105
Rock-wallaby, Allied 70
  Black-footed 74
  Brush-tailed 75
  Eastern Short-eared 76
  Godman's 72
  Herbert's 73
  Mareeba 74
  Monjon 71
  Nabarlek 72
  Proserpine 75
  Purple-necked 76
  Unadorned 73
  Western Short-eared 71
  Yellow-footed 77

Saccolaimus flaviventris 117
  saccolaimus 118
Sarcophilus harrisii 27
Scoteanax rueppellii 131
Scotorepens balstoni 131
  greyii 132
  orion 132
  sanborni 133
Sea-lion, Australian 140
  New Zealand 141
Seal, Southern Elephant 142
Setirostris eleryi 122

Setonix brachyurus 88
Sminthopsis bindi 37
  butleri 38
  crassicaudata 38
  dolichura 39
  douglasi 39
  gilberti 40
  hirtipes 40
  leucopus 41
  longicaudata 41
  macroura 42
  murina 42
  ooldea 43
  virginiae 43
  youngsoni 44
Sousa sahulensis 157
Spilocuscus nudicaudatus 63
Stenella coeruleoalba 158
  longirostris 158
Steno bredanensis 159
Sus scrofa 143
Swamp Buffalo 144

Tachyglossus aculeatus 20
Taphozous australis 118
  georgianus 119
  hilli 119
  troughtoni 120
Tarsipes rostratus 61
Tasmanian Devil 27
Thylogale billardierii 77
  stigmatica 78
  thetis 78
Tree-kangaroo, Bennett's 69
  Lumholtz's 70
Tree-rat, Black-footed 93
  Golden-backed 94
Trichosurus caninus 63
  cunninghami 63
  vulpecula 64
Tursiops aduncus 159
  truncatus 159

Uromys caudimaculatus 104

Vespadelus baverstocki 133
  caurinus 134
  darlingtoni 134
  finlaysoni 135
  pumilus 135

  regulus 136
  troughtoni 136
  vulturnus 137
Vombatus ursinus 50
Vulpes vulpes 138

Wallabia bicolor 87
Wallaby, Agile 81
  Antilopine 85
  Black-striped 81
  Bridled Nailtail 84
  Northern Nailtail 85
  Parma 83
  Red-necked 84
  Swamp 87
  Tammar 82
  Western Brush 82
  Whiptail 83
Wallaroo, Black 86
  Common 86
Whale, Antarctic Minky 152
  Blue 150
  Bryde's 150
  Dwarf Minky 151
  False Killer 157
  Fin 151
  Humpback 152
  Killer 156
  Long-finned Pilot 155
  Melon-headed 154
  Pygmy Right 148
  Pygmy Sperm 153
  Sei 149
  Short-finned Pilot 155
  Southern Right 149
  Sperm 153
Wombat, Bare-nosed 50
  Northern Hairy-nosed 49
  Southern Hairy-nosed 50
Wyulda squamicaudata 64

Zyzomys argurus 104
  pedunculatus 105
  woodwardi 105